Urban Farm Projects

MAKING THE MOST OF YOUR MONEY, SPACE, AND STUFF

by Kelly Wood

630.91732
WOD

I-5
PRESS

S0-AFI-307

Senior Editor: Jarelle S. Stein
Art Director: Cindy Kassebaum
Book Project Specialist: Karen Julian

I-5 PUBLISHING, LLC™
Chief Executive Officer: Mark Harris
Chief Financial Officer: Nicole Fabian
Vice President, Chief Content Officer: June Kikuchi
General Manager, I-5 Press: Christopher Reggio
Editorial Director, I-5 Press: Andrew DePrisco
Art Director, I-5 Press: Mary Ann Kahn
Digital General Manager: Melissa Kauffman
Production Director: Laurie Panaggio
Production Manager: Jessica Jaensch
Marketing Director: Lisa MacDonald

Library of Congress Cataloging-in-Publication Data
Wood, Kelly, 1969-
  Urban farm projects : making the most of your money, space, and stuff / by Kelly Wood.
      pages cm
  Includes index.
  ISBN 978-1-935484-78-3
1. Urban agriculture. 2. Quick and easy cooking. I. Title.
  S494.5.U72W66 2013
  635.9'77--dc23

                          2012041719

I-5 Publishing, LLC™
3 Burroughs, Irvine, CA 92618
www.facebook.com/i5press
www.i5publishing.com

Printed and bound in the China
13 14 15 16 17   1 3 5 7 9 8 6 4 2

Dedication

To my mother, whose beautiful Belgian hands will always be my image of absolute capability;
to my father, for teaching me that persistence and determination are omnipotent;
and to Roger, my love and best friend, for letting me.

Acknowledgments

Thank you to Andrew DePrisco for planting the seed of this book; Lisa Munniskma for referring me;
Karen Julian for all of her legwork and being my "Go-To Gal," even during her pregnancy;
and most of all to Amy Deputato for her diligence and hard work in editing my verbosity,
compiling everything into a cohesive unit, being constantly cheerful and upbeat and energetic
in the midst of her own toddler trio and the East Coast tribulations of the past year,
and, more than anything, putting up with my emotions and uncertainties throughout the process.
I would not have finished it if it were not for her. Many thanks to you all!

Contents

INTRODUCTION 6

SECTION I: IN YOUR KITCHEN

Project 1   Making Stock 12

Project 2   Homemade Pasta.................... 16

                Flat Noodles....................... 18

                Filled Pasta 20

Project 3   Homemade Dairy Products 22

                Chèvre Cheese 24

                Mozzarella Cheese 25

                Yogurt 26

Project 4   Condiments and Seasonings 28

                Ketchup............................ 30

                Mayonnaise 31

                Mustard 31

                Vanilla Extract.................... 32

Project 5   Baking Bread........................ 34

                Bettina's Recipe 36

                Jeff's "Sand Wedge" Bread.......... 37

                English Muffins 38

                Bagels............................. 39

Project 6   Small-Scale Beer Brewing............. 42

Project 7   Fruit Cordials...................... 46

Project 8   Homemade Mixers 50

                Tonic Concentrate 52

                Ginger Ale Concentrate............. 52

                Margarita Mix...................... 53

Project 9   Make Your Own Tea 54

Project 10 Canning and Pickling 60

Project 11 Drying Fruit 66

Project 12 Freezing Produce 70

Project 13 Root Cellaring...................... 74

SECTION II: FOR YOU AND YOUR HOME

Project 1   Cleaning Supplies 80

                No-Streak Window Cleaner 82

                Liquid Tub and Tile Cleaner 83

                Heavy-Grime Marble and
                Granite Cleaner 83

                Stainless Steel Polish.............. 83

Project 2   Skin-Care Products 84

                Lip Balm 86

                Baby's Bottom Cream................ 86

                Herbal Soap Balls 87

                Insect Repellent................... 87

Project 3	Make Your Own Candles	88
	Rolled Taper	90
	Poured Square Column	90
	Poured Round Column	91
Project 4	Dryer Bags and Sachets	92
	Lavender Dryer Bag	94
	Lavender Sachet Bag	95
	Cedar Sachet Bag	95

SECTION III: BACKYARD PROJECTS

Project 1	Build Your Own Rain Barrels	98
Project 2	Build a Bird Feeder	102
Project 3	Worm Bins	106
Project 4	Compost Piles	112
Project 5	Keeping Chickens	118
Project 6	Dairy in the Backyard: Goats	128
Project 7	Mason Bees	134

SECTION IV: OUTDOOR GARDENING

Project 1	Create a Garden Area	140
Project 2	Season Extension	146
Project 3	Plant a Pollinator Garden	152

Project 4	Edibles on Your Deck	156
Project 5	Make Your Own Planters	160
Project 6	Vertical Growing	166
Project 7	Raised Beds	170
Project 8	Growing in Tiers	174
	Tiers with Raised Beds	176
	Tiers with Planters	177
Project 9	Build a Trellis	178
Project 10	Irrigation Systems	182
Project 11	Gravity-Fed Watering Systems	190
Project 12	Self-Watering Planters	194
Project 13	Community Gardening	198

SECTION V: INDOOR GARDENING

Project 1	Windowsill Herb Gardens	204
Project 2	Setting Up Artificial Lighting	212
Project 3	Creating Miniature Topiaries	216

RESOURCES — 222
PHOTO CREDITS — 223
INDEX — 224

Introduction

*Receive a fish,
eat for a day.
Learn how to fish,
eat for a lifetime.*

—Variation on an ancient proverb

It seems that going hand in hand with the current technological revolution is widespread nostalgia for the past. There is renewed interest in self-sufficiency through traditional farm-style practices, and increasing demand for space in which to garden or raise livestock in close proximity to urban amenities. Many people realize that these practices can be done in smaller dwellings and on urban plots instead of solely on rural acreage. More of us are rethinking how and where food can be grown, leading to a surge in innovation and ingenuity. It's a movement toward simplification and getting back to the land while incorporating modern technology to facilitate the process. The challenge is how to optimize this on a functional, daily basis.

Modern life is too full—full of possessions, activities, news, and information. We have electronic screens in our homes, our offices, our cars, and even our pockets. Everywhere we turn, advertisements tout products that we "need" to make us happy and fulfilled. Every new item promises to streamline our lives, yet each one requires accessories and obligations—another power plug, another holder, another monthly fee. More and more, people are seeking less and less—fewer objects, fewer activities, less (or at least better) news, more

concise information. We want respite from the busy norm. Because "getting back to basics" differs from many of our current habits, it feels new. But many of these basics have been around for a long, long time.

There was a lot of life before automation. A few generations ago, the majority of American families lived on farms, raising almost all of their own food. Parents needed their children to help run the family farm; in turn, the grown children took over the farm from their aging parents. But far-off cities were growing, and the higher paying jobs, less physical toil, and greater excitement of city life began luring young people away from the farms.

In the early to mid-twentieth century, planners created new *sub-urbs* (from the Latin for "under" and "city"), which offered this younger population affordable housing with easy proximity to urban jobs. Families were able to live close to cities but still have modest houses and yards for kids and pets. The 1950s postwar American lifestyle, with newfound exuberance and affluence, embodied the quest for ease and leisure, resulting in people embracing "labor-saving" devices for

use in the home. Televisions became ubiquitous, and thus marketers discovered a ready audience for their sales pitches.

Despite the influx of tools designed to simplify cooking, there also was a new industry in grocery stores and ready-made foods. It became easier and more stylish to buy a package instead of assembling fresh ingredients to cook from scratch. Advertisers of modern conveniences strove to convince potential buyers that they'd wonder "how they ever lived without it!" There always seemed to be something newer and better to help homemakers or please the kids. Our collective discontent began growing.

Meanwhile, the migration away from family farms reduced the number of able bodies to inherit and work the land. As people aged out of farming and chose to sell their acreage, the fewer remaining farmers consolidated their holdings. These bigger farms needed to keep up with the demand for their products, and labor-saving methods were available to them also. Industrial practices emerged to facilitate running bigger operations, and

the family farm became quaint and old-fashioned, left behind in the drive toward large-scale food production.

Many studies have shown how environmental degradation can result from industrial production and processing. Certain practices contribute to the pollution of air, water, and soil, which are the foundations that food sources—plants and livestock—need to grow. When the foundations are unhealthy, the food quality suffers, and then we all suffer. Problems keep surfacing in our food supply; concerns have arisen about how and where food is being grown and raised, and where contaminants are coming from. The "epidemics" of obesity and diabetes demonstrate how the ingredients in processed foods can affect our bodies. More problems mean more need for solutions, and manufacturers are ready to fill this new need. Advertisers again tell us, "You won't know how you lived without it!" And so the cycle repeats.

Today, more people live in cities than in rural areas, but there is nostalgia for the perceived simplicity of country living, and a renewed appreciation for what small farms provide. People are beginning to understand

that "old-fashioned" doesn't mean "outmoded." Farm products—think fresh milk and eggs, whole grains, and homegrown produce—and the active processes of raising them are what many city dwellers seek. Some of this desire has come from increased scrutiny of our food-production systems, while many want to return to the way that previous generations ate—freshly prepared meals made from basic, wholesome ingredients.

I am a farmer and a full-time mother. Despite certain perceptions, being a stay-at-home parent does not bring limitless free time. For many of us, it ups the ante to do more because tasks such as laundry, cleaning, grocery shopping, and running errands are just not that mentally stimulating. I enjoy challenges, and I often encounter people who can't believe my activity level. They speculate that I must never sleep, and they don't know how they could fit the things I do into their own routines. In our ready-made consumer culture, they think doing things by hand is more difficult—why make it when you can buy it?

The pioneers and settlers who built our country had the same twenty-four-hour days that we have, but they had less time to work because they were beholden to natural cycles for light. They worked hard because they had only their physical strength and ingenuity to help them survive. The infrastructure that we take for granted didn't exist—they didn't have cars, paved roads, heavy machinery, electricity, phones, indoor plumbing and heating, convenience stores, garbage collectors, media, or the Internet. Nonetheless, they built sturdy structures, kept in contact with friends and family, drank clean water, prepared flavorful food and beverages, took medicine, kept animals, had social gatherings, grew gardens, preserved food, enjoyed music, cleaned their houses, washed their clothes, and bathed—just like us. And they took every Sunday off.

I have always been a curious person, seldom satisfied with the status quo. *How* and *why* are two of my favorite words. I constantly ask myself how things are made, or why I am buying something. There is little I like better than something to take apart or an opportunity to find a different solution. When a commercial or advertiser tells me that I'll wonder how I

SECTION I:
In Your Kitchen

ever lived without a certain product, I do exactly that—I wonder how people lived without it. If past generations got the job done without this gadget, how did they do it? Usually, they did just fine. Our great-grandparents grew up without the paraphernalia that we have, but they seem to have been healthier and happier. They made do with less because they knew how to do more. Many of the projects in this book are old practices that our generation has not been taught.

I think we should relearn how things used to be done. If I can buy an item in the store, then someone, somewhere, made it. How is it made? How did it used to be made? Is it something that I can make so I don't need to buy it or rely on a manufacturer for refills? Could I save time or money by making it? Would it be fun to try? When these questions motivate me to act, I inevitably learn something—either from the research, the process, or the mistakes made. The knowledge is out there to be had, and finding it and applying it are up to you.

I've compiled my years of giving in to curiosity in these pages. Although our predecessors worked incredibly hard, I found out that we can make many of the things they did, often more easily and in less time, resulting in healthier, simpler, more rewarding lives.

This book is intended for the average householder. Each project has an introductory page that gives an overview, including any specific skills required. Special equipment is also listed, but you may come up with your own ideas. Depending on how you do something, someone else's ideal tool might not work for you.

I try to operate on the assumption that you are not interested in accumulating a lot of specialized equipment and would rather use tools and devices that you already own—simplification means more "double-duty" items. Since my family relies on one income, I try to spend as little as possible and avoid superfluous purchases altogether. Don't buy something new for an activity you might not pursue, and be realistic. Remember, despite convincing advertisements, many "perfect solution" items are not perfect solutions at all—if they were, inventors would stop trying to improve upon them.

Some stores will allow you to "test-drive" an item, meaning that you can try it and return it if you aren't happy with it, but check their policies first. Or try borrowing—many people have items that you could use, and someone might even be interested in trying the project

with you, using the equipment together, and sharing the resulting yield. This is a great way to meet new people or get together with friends for a fun undertaking.

I have tried every project in this book. My results are not all pretty or perfect, but they are effective. My idea of an accomplishment may differ from yours, just as my definition of "functional" differs from my husband's perception of "nice to look at."

Feel free to adapt and adjust the parameters of any project—in other words, experiment! I am not the authority on these topics, and I've come up with my own methods through practice. Many of these projects are pretty forgiving, and if I haven't included exact measurements or recipes, you can safely assume that the process is fairly loose and play with it a bit. If it doesn't work, don't give up. Learn from it and try a different approach the next time. You'll work out what suits you best.

If you're particularly interested in a project or practice, by all means delve into it further. Resources are out there for anyone who wants to learn; I've included a few of my favorites with most of the projects.

You can find more resources online; there are countless websites on most of the topics discussed in this book, with many recipes and formulas to be shared. Also look for more in-depth books on your favorite topics—before you buy, check your local library for books that explain clearly and answer your questions. I have some books that are dog-eared from constant reference, and others I wish I hadn't wasted money on.

If you try something and decide it's not for you, at least you will have gained an appreciation for what goes into it. You'll be especially glad the next time you reach for the product on the shelf, thinking, "Thank goodness I don't have to make this myself!" It might even inspire you to find a local purveyor who has mastered the craft and makes an exquisite version and who might be willing to give you tips if you want to try the project again.

Humans are social creatures—we cannot do everything for ourselves, but there is a lot we can accomplish on our own. If we make time for what we believe is important, we can derive so much enjoyment from *doing* instead of *having*. We can simplify and actively learn simultaneously. When we work hard to achieve something, the outcome is even more fulfilling. We all have the potential for this kind of satisfaction. Give it a try.

GRAVES,
MAGGIE

Pickup by:
07/23/2020

3243
Urban farm projects : making the most of your

money, space, and stuff
21487003185646

Making Stock

Stock, a culinary staple made from ingredients that you already have on hand, is an easy project—one that creates a base for soup, stew, roux, jus, gravy, or anything else that calls for broth. You simply let the ingredients sit and simmer until the house smells good, and then strain the liquid. Once I saw how effortless the process was—and how much better the results were—I chided myself for having bought the commercially made version for so long. When you start making your own stock, you'll resort to the store-bought variety only in emergencies (such as the imminent arrival of unexpected dinner guests!).

Why would you want to do this?
Homemade stock is much healthier than the packaged product, and it's very easy. Making stock is a great way to get more than one meal out of leftovers you'd otherwise throw out or compost.

Why wouldn't you want to do this?
I can't think of a reason.

Is there an easier way?
It's hard to get much easier than making stock.

How is this different from the store-bought version?
Your stock will taste fresher and be much lower in sodium. It may have a residue at the bottom, but that's just the leftover bits of whatever you cooked. When refrigerated, meat stock can become gelatinous, but it will liquefy again when heated.

Cost comparison:
There's no additional cost for homemade stock because it's made from ingredients that you've already purchased for meals.

Skills needed:
Homemade stock is hard to mess up. If you can boil water, you should have no problem with this project.

Further refinements:
You can experiment with many seasonings and flavors. Also, after the stock is strained, you can return it to the pot and cook it again, uncovered, to reduce it and make it more concentrated. You then freeze the concentrated liquid in ice-cube trays and use the frozen cubes similar to the way you'd use bouillon cubes.

Some trimmings, herbs, and water are essentially all you need for delicious homemade stock.

There are essentially two main varieties of stock: meat and vegetable. Although in our kitchen, we most often use chicken, beef, and vegetable stock, I've noticed more exotic versions—such as clam and mushroom—on grocery-store shelves. Stock is a simple cooked infusion of flavors that is used as a base for, or to enhance, other foods. You make stock with what is left over from meal preparation—what you otherwise would typically throw out.

Homemade stock is healthy. As you cook vegetable stock, the vitamins and nutrients from the leaves, peels, and ends of vegetables are drawn out into the broth. Like-

wise with meat stock—the extended cooking time releases the valuable nutrition in the bones and marrow, presenting the latter in a palatable form. My kids would never consent to eating marrow, but they like foods made with meat stock, and they get the benefits of the marrow's high iron content while enjoying what they're eating.

Our friends Jeff and Cindi taught us to put the vegetable trimmings from our daily meal preparation in a freezer container—I include meat bones in ours—until we've collected enough to make a batch of stock. If you use whole fresh vegetables, the cooked parts left after your stock is prepared can go directly into a soup.

Materials/Ingredients:

- ☐ Medium to large stockpot
- ☐ Colander or strainer
- ☐ Optional: Cheesecloth
- ☐ Optional: For freezing—canning jars or freezer containers; for refrigerating—sealable rigid containers

Vegetable Stock: Vegetables, whole or trimmed, any or all of the following:
- ☐ Carrots, including roots, ends, and peels
- ☐ Onions, including bulbs, skins, and peels
- ☐ Garlic, including cloves and skins
- ☐ Celery, entire stalks with leaves
- ☐ Potatoes, including tubers and peels
- ☐ Scallions (wash root ends if still attached)
- ☐ Shallots, including bulbs, skins, and peels
- ☐ Herbs (bay leaves, rosemary, sage, parsley, basil, oregano, thyme), fresh or dried, complete branches or leaves only
- ☐ Seasonings to taste

Meat Stock: Use the ingredients list for vegetable stock, plus anything remaining from the carcass, such as:
- ☐ Bones
- ☐ Skin (from poultry)
- ☐ Meat (cooked or raw)

Step 1: Put all of the ingredients except seasonings into the stockpot. Add enough water to cover everything (if you've steamed your vegetables, you can save the water and use that) and bring to a boil.

Soup made with homemade stock is full of healthy, hearty goodness.

Step 2: Once the contents are boiling, turn the heat down to a simmer and let the stock cook slowly, covered. While it is simmering, add salt and other seasonings to taste.

Step 3: Cook vegetable stock for at least an hour; cook meat stock longer, especially if you're using raw scraps. Stock is one of the few dishes that I feel comfortable leaving on the stove for hours; it's one of those rare dishes that cannot be overcooked. The longer it's on the stove, the more the flavor is enhanced.

Step 4: Remove the stock from the heat. Strain it through a mesh strainer or colander to get rid of any bones or any big chunks of vegetables or meat, and reserve the liquid. For a clearer broth, strain the liquid a second time through cheesecloth.

Step 5: If the strained vegetable and meat pieces are usable, chop them up for soup; otherwise, compost or throw away the vegetable matter, and throw away the meat and bones.

Use or store the stock right away. You can refrigerate it in a glass jar or plastic storage container for up to a week. For longer storage, pressure-can the liquid in jars (refer to Project 10 in this section, Canning and Pickling) or freeze it, leaving room in the containers for expansion.

Voilà! You've just taken what would have otherwise been garbage and turned it into something delicious that can be used in many more meals.

Homemade Pasta

The Italians have been making pasta for countless generations. It doesn't take as long as you might think, and it makes a huge difference in the taste of your meal. Store-bought fresh pasta is a high-end delicacy, but you can save yourself the trip to the market.

Why would you want to do this?
Fresh pasta cooks faster and is superior to its dried counterpart in taste and texture. Making pasta is a great medium for creativity in the kitchen once you're used to making the dough.

Why wouldn't you want to do this?
It takes more time than grabbing a bag or box of store-bought noodles from the pantry.

Is there an easier way?
A pasta rolling machine makes the process much more consistent. It usually comes with cutters for certain types of pasta. The machine and any additional attachments you may want are expensive.

How is this different from the store-bought version?
Homemade pasta has a different "mouthfeel" and taste than dried pasta, and you need to cook it carefully so it doesn't become overcooked and gummy (the recommendation of cooking pasta *al dente* comes from cooking fresh noodles).

Cost comparison:
This recipe makes 1 pound of pasta (enough for our family of four) and the ingredients cost less than 1 pound of fresh packaged pasta.

Skills needed:
A comfort level with cooking from scratch or a willingness to experiment and deal with a few initial failures.

Learn more about it:
Making Artisan Pasta (Quarry Books, 2012) by Aliza Green offers inventive tips and tricks.

Crack the eggs into the top of the "volcano."

I've got to confess that pasta-making is my husband's territory. I still occasionally rely on dried store-bought noodles, but when I'm putting together a special dinner or just don't want to run to the store, I make the pasta from scratch and am always impressed by what it adds to the meal. With practice, you'll discover how easy the process can be and how tasty the results are.

Traditional pasta is made from hard-grained wheat called *durum wheat*. The flour is a semicoarse grind called *semolina*; finer grinds are called *durum flour*. (Under Italian law, dried pasta must be made from semolina.) When making pasta from scratch, you can use either semolina or plain white flour—the latter is what we call *bread flour*—or a mixture of both. My husband uses whatever is on hand.

Semolina-only noodles will be chewy, more like a packaged dried pasta when cooked. Pasta made with bread flour alone can be gummy if overcooked, but when cooked properly, the texture of this pasta is more delicate than that of pasta made with semolina only.

Flat Noodles

• •

Materials/Ingredients:
- ☐ Mixing bowl
- ☐ Rolling pin (or wine bottle)
- ☐ Knife or pizza/pasta cutter
- ☐ Optional: Mixer with dough hook
- ☐ 3 large eggs
- ☐ 2 cups flour
- ☐ 2 tsp olive oil
- ☐ Water as needed
- ☐ ¼ tsp salt

• •

After the dough has rested, you can start rolling it out on your work surface.

Keep rolling the dough until you've gotten it flat, thin, and even.

Finished flat noodles, ready to be cooked or dried.

Step 1: Mix the flour and salt together. Make a mound of these ingredients on a clean countertop or in a bowl and then hollow out the top, like a volcano.

Step 2: Put the eggs and oil into the hollow and then mix them into the flour mixture; it is best to do this with your hands, but you can also use a mixer with a dough hook. Keep mixing until a ball develops and all of the dry crumbs on your fingers are incorporated into the dough. If the dough gets too dry, sprinkle it with a bit of water.

Step 3: Knead the dough for five to ten minutes. Shape it into a ball again and let it sit for ten to thirty minutes, covered or wrapped in plastic, to allow the glutens to relax (which will help the dough roll out more easily).

Step 4: Roll out the dough on the countertop to an even thickness, as flat and thin as possible, and then fold it over and over onto itself (alternating top to bottom and side to side) until it looks like a nifty little folded handkerchief.

Step 5: Roll the dough out again and then fold it again in the same way; repeat this process several times until the pasta dough appears to be an even color and consistency.

Step 6: Roll out the dough one more time, but do not fold it; leave it very thin and flat. Use a knife or a pasta or pizza cutter to cut the dough into noodles of your desired size and shape (I used to use our kids' Play-Doh cutters with curvy edges).

Step 7: Set the noodles aside after you cut them. When we make long, thin noodles, we drape them carefully across the backs of our kitchen chairs—and the banisters, and towel bars, and counter edges!

Either use the pasta noodles right away by cooking them rapidly in boiling water or leave them out to dry completely (they will be very brittle) and store them in a plastic bag or container.

Eccolo—homemade pasta! Enjoy!

Filled Pasta

Once you've mastered rolling out sheets of pasta, you can make filled pasta, too. We've hosted parties where each couple brought a filling, we provided the sheets of pasta, and everyone had a boisterous evening of creating and cooking delicious filled pasta. Our guests got to take home big sealed plastic bags of pasta for their freezers.

• •

Materials/Ingredients:

☐ Ravioli tray or ice-cube tray
☐ Small spoon
☐ Paintbrush or pastry brush
☐ Glass of water
☐ Pasta or pizza cutter
☐ Rolling pin or wine bottle
☐ Parchment paper
☐ Cookie tray
☐ Optional: Freezer bags or containers
☐ Fresh pasta sheets
☐ Filling of your choice (see Basic Ravioli Filling sidebar)

• •

Ravioli

We've tried many different ravioli-making tools over the years and ultimately decided that the device resembling an ice-cube tray is our favorite. You can use a regular ice-cube tray to similar effect, or you can separate the ravioli with your pizza or pasta cutter; both methods are explained here. Either way, cut your pasta sheets to the same size, as one will lie on top of another. If using a tray, cut your pasta sheets so that they are slightly larger than the tray.

Put the completed ravioli on a parchment-lined cookie sheet to be frozen or cooked. If storing, leave the ravioli

Basic Ravioli Filling

☐ 15-ounce container of ricotta cheese
☐ 1 egg
☐ $1/8$ cup Parmesan cheese, finely grated
☐ $1/2$ tsp fresh Italian parsley, chopped
☐ Cinnamon, pinch

Mix all ingredients together until well combined and smooth. Use a dollop in each ravioli pillow.

on the cookie sheet in the freezer until they are semi-frozen before putting them in a container or freezer bag (they should be firm enough to be stacked without getting mashed). To use them right away, cook them by dropping them individually into a wide kettle of boiling water.

The Tray Method:
Step 1: Lay a single sheet of pasta on top of the tray, and gently make indentations into the openings below. Be careful not to tear the sheet as you press down.

Step 2: Scoop a scant spoonful of filling into each indentation.

Tortellini are hand-folded into the "belly-button" shape.

Step 3: Paint lines of water where the pasta is touching the edges of the tray—between the indentations and on the outside edges.

Step 4: Lay the second sheet over the first, matching up the outer edges. Roll firmly over the top of the pasta with a rolling pin (or wine bottle), sealing the edges.

Step 5: Lift the sealed pasta sheets out of the tray, and, if necessary, cut between each ravioli with a knife or wheel to separate them.

The Countertop or Tabletop Method:
Step 1: Dust your work surface lightly with semolina, and lay out one sheet of pasta.

Step 2: Place scant spoonfuls of filling on the pasta, spacing them out evenly and leaving at least one or two finger widths between each scoop.

Step 3: With the brush, paint water lines between all of the dollops of filling.

Step 4: Gently lay the second sheet of pasta on top of the first, lining up the outer edges. Carefully press the pasta down between the dollops; you want as little air in the pockets of filling as possible.

Step 5: Take your cutter and firmly roll between each of the dollops along the wet lines you painted. Be sure to cut all the way through both sheets. Although some cutters will seal the edges as they cut, we've found that it still helps to pinch the edges of each ravioli to make sure that they are sealed on all sides. If the ravioli are not sealed, they will spill their contents into the cooking water.

Tortellini

Tortellini are ring-shaped, filled noodles whose shape, legend has it, was inspired by Venus's belly button. Whether the Roman goddess of love had anything to do with it or not, this pasta is well loved and worth the time and effort involved in creating your own. Each noodle is made from a round or square piece of pasta, and the navel shape is formed after the noodle is filled and sealed.

A cutter separates the ravioli and helps seal the edges.

Step 1: Decide whether you want circular or square tortellini. To cut circles out of the flat pasta sheets, you can use a clean tuna can, a cookie cutter, or a biscuit cutter—all work well. For squares, use a knife to cut the pasta into pieces of equal size.

Step 2: To fill the tortellini, center the filling on half of the circle or square, and then fold it in half (for the squares, fold diagonally into triangles). Pinch the edges to seal them.

Step 3: When your half-circles or triangles are complete, the filling bulge should be against the fold and centered between the sealed edges.

Step 4: To fold a sealed pocket into the navel shape, gently place the fingernail side of your index finger against the filling bulge. With your other hand, carefully wrap the corners around your finger, pinching the two ends together firmly with your thumb.

Step 5: Slide the folded pocket off of your finger and set it on a parchment-lined tray. (The square tortellini can look like little pointy hats when folded; you can fold the points down to minimize this effect, if desired.)

Freeze or cook as described for the ravioli.

Homemade Dairy Products

The process of making your own milk-based products is a lot of fun and a great learning experience. I encourage everyone to try it at least once. If you don't like the taste of what you've whipped up, don't throw in the towel right away. Give it another try—or, better yet, several more tries—before you concede defeat. There are so many subtleties and nuances to cheeses and yogurts that making them can be

Why would you want to do this?
It's fun, it's educational, and you'll impress yourself and your friends with the tasty results.

Why wouldn't you want to do this?
You are lactose intolerant, you don't like cheese or yogurt, you don't want to buy extra gallons of milk, or you don't have access to fresh farm milk.

Is there an easier way?
No, but it only gets easier the more you do it!

How is this different from the store-bought version?
Homemade versions have a richer taste—*truer* is how I describe it, and I believe it is because they are fresher than manufactured products.

Cost comparison:
For less than the cost of buying these products from the grocery store, you can buy the materials and ingredients and have a larger yield.

Skills needed:
Basic cooking skills and patience.

Further refinements/learn more about it:
Keeping notes about your successes and failures will help you refine your technique. Learn from Ricki Carroll (the "Cheese Queen") and her book, *Home Cheese Making* (Storey, 2002), the source of the cheese recipes that follow. Her website, www.cheesemaking.com, offers plenty of advice as well as materials and ingredients. *Cheese It!* (I-5 Press, 2012) by Cole Dawson is a detailed, descriptive guide geared toward novices.

an addictive hobby once you get the hang of it.

I am fortunate to be able to keep dairy goats, so I have a constant supply of fresh milk. I have to figure out ways to use it because fresh milk doesn't keep as long as processed milk does, and the goats give us more than we can drink. All of the projects discussed here can also be done with store-bought milk.

You'll have to decide if the price and process of purchasing the materials and ingredients are worth it for you (although the cost is still less than grocery-store prices), but making your own dairy products is fun, and the results are tasty.

Materials for Each Project:
- ☐ **Large (8-quart or larger) nonreactive stockpot (stainless steel or enameled)**
- ☐ **Measuring cups and spoons**
- ☐ **Slotted spoon**
- ☐ **Long-handled wooden or plastic utensil for stirring milk**
- ☐ **Thermometer that registers from 40 to 212 degrees Fahrenheit (boiling)**

As the curds solidify, they begin to pull away from the sides of the stockpot.

Finished chèvre, coated with herbs and spices.

Chèvre Cheese

I make chèvre on a weekly basis—it is one of the easiest tasks I have, and it gives us (and my customers) a steady supply of a versatile cheese. This recipe is shared with us courtesy of Ricki Carroll from her book *Home Cheese Making* (Storey, 2002).

Ingredients/Additional Materials:
- ☐ **1 gallon whole milk (goat's milk is best, but cow's milk is fine)**
- ☐ **1 packet chèvre starter**
- ☐ **Colander or fine strainer**
- ☐ **Optional: Cheese mold with drainage holes**

Step 1: In a large kettle, heat the milk to 86 degrees Fahrenheit.

Step 2: Turn off the heat and add the starter, stirring gently until it is well incorporated. Move the kettle to a back burner and let it sit, covered, for twenty-four hours.

Step 3: Uncover the kettle. You'll see a solid white mass in the midst of milky-looking water because the milk has separated into *curds* (the solid proteins) and *whey* (the liquid remaining after the curds coagulate). Pour the contents of the pan into a fine strainer or colander (or a cheese mold with holes), making sure that all of the solid curds end up in the colander/strainer or mold. Optional: Scoop the curds first into a large bowl and mix them with herbs and seasonings, and then pour them into the colander/strainer or mold for draining and shaping.

Step 4: Pour any last bits of curd or whey out over the curds, letting the whey strain through. Let the curds sit in the colander/strainer or cheese mold until well drained, from twelve to twenty-four hours. Longer draining time will yield a drier result.

Note: If you drain the cheese in a cylindrical cheese mold, it will form a tube shape; when the cheese is fully drained, you can roll it in flavorings such as crushed dried herbs, coarse ground pepper, or sesame seeds.

Store the finished product in a covered container in the refrigerator; eat it within a week.

Mozzarella Cheese

Ricki Carroll has also shared her mozzarella recipe from her book *Home Cheese Making* (Storey, 2002).

● ●

Ingredients/Additional Materials:

☐ 1 gallon whole goat's or cow's milk (whole milk is best, though 2 percent will work; pasteurized milk is fine, but do not use ultra-pasteurized, which does not allow the cheese to set properly)

☐ ¼ cup unchlorinated water

☐ Food-grade citric acid (available from a brewer's supply or where cheese-making supplies are sold)

☐ Rennet*, in liquid or tablet form (available at some grocery stores or from cheese-making sources; vegetarian rennet is also available)

☐ Optional: Cheese salt (such as kosher salt), to taste

☐ Microwaveable glass bowl

*Junket rennet comes in tablet form and is weaker than real cheese rennet. If using junket rennet, use a whole tablet instead of a quarter where specified.

● ●

Step 1: Dissolve 1½ teaspoons citric acid in ½ cup room-temperature unchlorinated water. Set aside.

Step 2: Dissolve ¼ teaspoon liquid rennet or ¼ of a rennet tablet in ¼ cup cool unchlorinated water. Set aside.

Step 3: Put the thermometer into the stockpot and pour in the milk. At 55 degrees Fahrenheit, add the citric-acid solution and stir gently but thoroughly. Heat the milk to 90 degrees over medium-low heat—it will begin to curdle.

Step 4: Gently stir in the diluted rennet solution, mixing thoroughly but carefully. The milk will begin to coagulate.

Step 5: Continue heating the newly formed mass of curds to 100–105 degrees and then turn off the heat.

As you heat the milk and citric acid, the mixture will start to curdle.

Finished mozzarella, braided.

The curds should be pulling away from the sides of the pot, and you will see what is called a "clean break" if you insert the stirring utensil (meaning that the curds split cleanly around the utensil). If the curds look like thick yogurt and the whey is clear, you are ready to scoop out the curds with the slotted spoon. If the whey is still milky, wait a few more minutes.

Step 6: Using the slotted spoon, scoop the curds out of the pot into the microwaveable bowl, letting the whey drain into the pot as much as you can with each scoop.

Step 7: Press the curds gently with your hands, pouring off as much of the whey as possible, either into the sink or back into the pot.

Step 8: Microwave the curds for one minute on high. Remove from the microwave and gently fold the cheese over and over onto itself—like kneading bread—with a spoon or with your hands. You want to distribute the heat evenly throughout the cheese to get it to the stretchy taffy stage (see Step 9), where it is smooth and elastic. Knead quickly so that you do not burn your hands.

Step 9: Microwave the cheese on high two more times for 35 seconds each time, kneading again after each heating. When the cheese is shiny and stretches like taffy, it is done. If the curds break instead of stretch after the third heating, they are still too cool; microwave them for another 35 seconds.

Note: If desired, add salt after the second heating. I start with 2 teaspoons but often add more to taste.

Step 10: When the cheese is done, roll it into balls, stretch it into sticks, or braid it while it's still warm; it gets less malleable as it cools down. Drop the shaped pieces into a bowl of ice water to cool them quickly; this will also produce a consistent texture. The cheese is best to eat warm but is firmer when chilled; however, I think that the texture is marvelous either way.

Store the mozzarella in the refrigerator in plain or lightly salted water ("brine") for up to a week. Ours seldom lasts that long because it's so tasty.

Yogurt

I have a yogurt maker, which is just a machine that holds the yogurt at a consistent temperature for the required time for it to set properly. It's nothing fancy, and yogurt can be made without one. If you don't use a thickener, the yogurt can be fairly runny, but it's still good to pour over granola or add to a smoothie. I just add fresh fruit and a bit of honey to my yogurt when it is done setting, and it suits my family fine.

Ingredients:
☐ **1 quart milk, any type**
☐ **¼ cup dry milk powder (as a thickener)**
☐ **1 packet yogurt starter or 2 Tbsp yogurt with live cultures**
☐ **Optional: 1 Tbsp thickener, such as carageenan, pectin, or gelatin (as a substitute for or in combination with the dry milk powder)**

Step 1: Combine the milk, milk powder (if using), and thickener (if using). Heat the mixture to 180 degrees Fahrenheit slowly and carefully. You don't want it to boil, and stirring it constantly should keep it from scalding.

Watch your temperatures carefully. Use a thermometer for both the heating and cooling steps.

Step 2: Let the milk cool to 116 degrees Fahrenheit. Add the starter and mix well.

Step 3: Keep covered at 116 degrees for at least six hours or until the yogurt has set to the consistency of thick cream. This is where a yogurt maker is helpful, but if you don't have one, there are various other methods. Some people preheat the oven to 120 degrees Fahrenheit, turn off the heat, and put the yogurt in overnight with the oven door closed. Some people use a slow cooker on low or a camping cooler with open jars of hot water surrounding the milk. Refrigerate and serve cold. The yogurt keeps for up to two weeks in the refrigerator.

Starter

The finished yogurt contains starter, so you can save 2 tablespoons of the yogurt to make your next batch. Each time you "reuse" the starter, it gets a little weaker, so when a batch of yogurt doesn't set at all, you know that the starter has stopped working. Throw out the old batch and open a new packet of starter powder.

Homemade yogurt topped with berries makes a healthy and tasty breakfast or snack.

Condiments and Seasonings

Store-bought condiments and flavorings are good examples of how we spend more money for the convenience of having something made and packaged for us. Flavors and simplicity aside, the price difference of making these products yourself is enough to encourage you to try it.

Why would you want to do this?

If you like to cook, these are quick and easy products to experiment with.

Why wouldn't you want to do this?

You don't want to clean out your blender (that was my excuse for a while!).

How is this different from the store-bought version?

The textures can be a little different; for example, homemade mayonnaise is oilier, and homemade ketchup separates a bit. You may also notice a taste difference. I like the challenge of trying to figure out which flavors are missing from my versions and seeing how close I can get to the mass-produced stuff, but often some of the tastes come from preservatives or chemicals, which you don't have in the homemade products.

Cost comparison:

Each of these recipes results in a lower cost per ounce than store-bought varieties; the biggest savings is with the vanilla extract.

Skills needed:

Basic cooking skills.

Homemade ketchup is a fresh take on a condiment that's a staple in many homes. To can your ketchup in jars, as shown here, refer to the Canning and Pickling project (Project 10) in this section.

We quickly and routinely deplete condiments in our house. Inevitably, a scant spoonful is all that the kids leave in the container. My eldest sister taught me years ago how easy condiments are to make, and the homemade versions are much healthier because they don't contain high-fructose corn syrup and many of the preservatives needed to keep mass-produced products fresh on grocery-store shelves.

Homemade condiments will spoil faster than store-bought versions, but they are easy to make in smaller quantities. I can honestly say that making your own can save you both money and time—I once raced my husband and made a batch of mayonnaise before he could get to the nearest store and back.

I don't think that the homemade versions taste too different from store-bought. I refill the brand-name jars, and our teenage son hasn't noticed (yet). And emulsification is fun to watch: it's chemistry in action.

Materials for Each Project:

- ☐ Food processor or blender
- ☐ Measuring spoons and cups
- ☐ Rubber scrapers
- ☐ Storage containers and lids

Ketchup

You can make ketchup from cooked, pureed fresh tomatoes (remove seeds, as they impart a bitter flavor) or from canned tomato puree. This recipe yields about 5 cups.

Ingredients:

- ☐ 4 cups fresh tomato puree (or a 28-ounce can)
- ☐ 1 medium yellow onion, peeled and chopped
- ☐ 1 clove garlic, peeled and crushed
- ☐ 2 Tbsp brown sugar (dark gives a richer flavor)
- ☐ ½ cup apple cider vinegar
- ☐ 1 cup water
- ☐ Pinch or more of each of following, according to your taste: ground allspice, finely ground black pepper, cayenne pepper, celery salt, ground cinnamon, ground cloves, dry mustard, ground ginger, kosher salt

Step 1: Put ingredients into a blender or food processor.

Step 2: Puree until completely smooth.

Store in the refrigerator in a squeeze bottle. If the mixture separates, shake well before squeezing out.

Mayonnaise

When I called my mother for her mayonnaise recipe, she was able to recite it right off the top of her head. I have vivid recollections of watching her make mayonnaise in her food processor and being fascinated. I marveled at the thin stream of oil slowly emulsifying the egg, completely transforming both ingredients. The recipe that follows yields approximately 1 cup.

• •

Ingredients:

- ☐ 1 egg
- ☐ 1 cup oil
- ☐ 1 tsp dry mustard powder
- ☐ ½ tsp salt
- ☐ ¼ tsp white pepper (you can use black pepper, but it makes specks in the mayo)
- ☐ 1½ tsp white vinegar
- ☐ 1–2 Tbsp lemon juice

• •

Step 1: Crack the egg into the blender or food processor and then add the mustard powder, salt, and pepper.

Step 2: Start the blender or food processor. As it runs, slowly drizzle in the oil in a thin, steady stream (a blending device usually has a trough in the lid with a small hole for this purpose).

Step 3: When all of the oil has been added and the mixture is emulsified, stir in the vinegar and lemon juice—you may need to do this by hand—until everything is completely combined.

Store the mayonnaise in a jar in the refrigerator; it keeps for up to two weeks.

Mustard

This recipe is simple yet full of possibilities for getting creative. You can vary the color of the mustard seeds to impart different tastes, and you can vary the grind to create different textures. You can change the proportion of vinegar to water and add different seasonings for a wide variety of flavors. This basic version has an initial spicy bite.

TOP: Eggs and oil are the main ingredients in mayonnaise.
BOTTOM: In a variation on the method presented in the recipe, you can use a hand mixer while slowly adding the oil.

Once you've tried the basic recipe, experiment with flavor by using different types of mustard seeds.

Ingredients/Additional Materials:
- ☐ ⅓ cup mustard seeds (brown or yellow)
- ☐ ¼ cup white vinegar
- ☐ 2 Tbsp water
- ☐ ¾ Tbsp sugar
- ☐ ¼ tsp salt
- ☐ Spice grinder or coffee grinder

Step 1: Prepare your seeds according to the finished texture you want. For smoother mustard, grind the raw seeds in a spice grinder (for a rough grind) or coffee grinder (for a fine powder). Raw seeds are chewy, so toast them before grinding to make them more brittle and easier to grind. For coarser mustard, keep the seeds intact but soften them by soaking them overnight in vinegar and water.

Step 2: Combine the prepared seeds with the remaining ingredients in a bowl or a food processor. Refrigerate for twenty-four hours, stirring occasionally (the mixture will thicken).

Store the mustard in a jar or a small lidded crock in the refrigerator. It will keep for at least two weeks.

Vanilla Extract

We've made our own vanilla extract for so long that I'd forgotten how tiny and expensive the store-bought bottles are. Whole vanilla beans are pricey—individual beans can cost more than a dollar each, and you'll use six to ten beans for a pint of liquor. Let's figure on $1 per bean for ten beans and around $10 for the bourbon, so a pint of homemade vanilla extract is going to cost you about $20. Because most recipes that call for vanilla extract require only a small amount, a pint will last you for a long time, and it's still much cheaper per ounce than store-bought vanilla. We just keep topping our jar off with bourbon.

Ingredients/Additional Materials:
- ☐ 10 whole vanilla beans
- ☐ 1 pint bourbon (choose an inexpensive brand, but not rotgut)
- ☐ Optional: Pint jar

It's normal for your homemade vanilla to appear a bit cloudy.

Step 4: Store the bottle in a dark place for five to six months, shaking gently once a week or so to blend and infuse.

Use the extract directly from the bottle; no filtering is necessary.

Step 1: Slice down the length of each bean, through just one side, leaving the ends intact and opening it up to expose the seeds.

Step 2: Stand the beans up in a jar or bottle that will hold a pint of liquid. Cover the beans with the bourbon. You can simply put the beans into the bourbon bottle if you pour out a little bit to compensate for the beans' volume.

Step 3: Cap the bottle and shake it gently.

Spices

Because so many spices are exotic, they can be expensive, especially if you buy them prepackaged in little bottles or jars. Even buying in bulk, by the ounce or pound, can add up if you use a lot of a particular spice or spices. If this is the case, consider buying the seeds or nuts of your favorite spices and grinding them yourself. You'll get more ground spice for your money. Grind only a little at a time; the flavor starts to deteriorate once the whole spice is ground. Unless otherwise noted, use a coffee grinder:

- Coriander (the very prolific seed of cilantro)
- Cloves
- Allspice
- Nutmeg (special grinders store whole nutmegs inside)
- Mace (the "web" of veining from the outside of a nutmeg)
- Cayenne pepper
- Ginger (peel, store in freezer, and grate frozen root as needed)
- Cumin
- Mustard

Ground (RIGHT) and dried (LEFT) cayenne pepper.

Baking Bread

It makes so much sense to bake your own bread. Everything about homemade bread is better: it is much healthier, cheaper, and tastier than store-bought; it makes the house smell wonderful; and, believe it or not, once you get the hang of it, making bread can feel easier than going to the market.

Why would you want to do this?
You want to eat better and exercise your creative side in the kitchen.

Why wouldn't you want to do this?
You don't eat bread, or you don't enjoy baking.

Is there an easier way?
Bread machines have many benefits (they do all of the work for you, and you can wake up to freshly baked bread) but just as many drawbacks (they are expensive; they are big; in my family's opinion, the crust isn't great; you can waste ingredients on failed batches and not find out until it's too late). The more you bake bread by hand, the more you'll see how easy it can be and the more you'll want to experiment.

Cost comparison:
As with many homemade food products, the materials cost less and yield more than the packaged versions.

Skills needed:
No special skills are needed. My mom taught me how to bake bread when I was still in grade school.

Learn more about it:
Knead It! (I-5 Press, 2013) by Jane Barton Griffith is a comprehensive guide to baking artisan bread, enhanced by interviews with and tips from accomplished bakers. *The Bread Baker's Apprentice* (Ten Speed Press, 2001) by Peter Reinhart is one of the better books to help you advance and perfect your bread-baking skills. Check local community colleges, other adult-education centers, or cooking schools for bread-baking classes.

Materials for Each Project:

- ☐ Mixing bowls
- ☐ Measuring cups and spoons
- ☐ Bowl scraper
- ☐ Mixing spoon
- ☐ Lightweight dishcloth
- ☐ Baking sheets or loaf pans
- ☐ Cooling racks

Bettina's Recipe for Super-Simple Rolls or Bread

If you're grinding your own flour (see Grinding Whole Grains into Flour sidebar on page 41) for baking bread, this is an easy recipe to start with. It uses half freshly ground whole-wheat flour and half unbleached all-purpose store-bought flour. As you get more comfortable with grinding and baking with your own flour, you can start to use less all-purpose flour and add more whole-grain flour (of different types) and groats, seeds, nuts, herbs, and dried fruits.

Ingredients/Additional Materials:

- ☐ 3½ cups whole-wheat flour or whole-spelt flour
- ☐ 3½ cups unbleached all-purpose flour
- ☐ 1 Tbsp salt
- ☐ 3¾ cups lukewarm water
- ☐ 2 Tbsp oil (canola, corn, or olive)
- ☐ 1½ Tbsp active dry yeast
- ☐ 1 tsp sugar (I use organic sugar)
- ☐ Parchment paper

The beautiful golden-brown color of fresh-baked bread.

Step 1: Dissolve the yeast and sugar in the water and then let it stand while you measure the dry ingredients into a bowl.

Step 2: Mix the dry and wet ingredients together. Knead the dough by hand for about five to ten minutes and then let the dough rise in the bowl for thirty minutes.

Step 3: Either form the dough into rolls and put them on a parchment-covered baking tray or make a loaf and put it in a loaf pan.

Bettina's Bread-Baking Hints

- You don't need a thermometer for the water—if it feels warm, not hot, then it's fine. Water above 110 degrees Fahrenheit will kill the yeast and will feel hot to the touch.
- If you like a crisper crust, brush the rolls or loaf with water or salted water before you put them in the oven or halfway through the baking time (or both)—experiment!
- If you don't want the bread to split naturally during baking, cut the top with a sharp knife or scissors after forming the rolls or loaf. I usually make one cut in the middle of each roll and three cuts across a loaf, but variations to suit your personal preference are fine.
- When you're kneading the dough, if it sticks to the bowl, add more flour. If you see dry flour in the bowl, add more water.
- You really don't need to let a "light" dough (such as the one in Bettina's recipe) have multiple risings. If you use more whole-grain flour or groats or whole kernels, then you would knead it and let it rise a second time before putting it in the pan.
- If you add groats or oats, soak them before you use them. Bring water or milk to a boil, pour it over the grains so that they're just covered, and let them stand for at least thirty minutes. These grains add liquid to the flour-yeast mixture, so you likely won't need additional liquid.

Step 4: Cover the dough with a lightweight dishcloth and let it rise in a warm place for twenty minutes (I put it on top of a heater vent).

Step 5: Put your rolls or bread in the oven and turn the oven to 400 degrees Fahrenheit (do not preheat; you want to give the dough a little more time to rise). Once the oven reaches 400 degrees, bake the rolls for about twelve minutes or the loaf for about twenty-five minutes, rotating midway if your oven has hot spots.

Step 6: Remove the loaf or rolls from the oven when they are lightly browned and sound hollow when tapped underneath.

Jeff's "Sand Wedge" Bread

This recipe comes from our friend Jeff, who has elevated the recipe and process to an art form. He got a scale for Christmas a few years ago and found that weighing ingredients (as Europeans do) makes all the difference in producing consistent results with any baked good. He can attest to the store-bought-like perfection and texture of these loaves, and he's been through hundreds of practice runs and adjustments to get the recipe to this point. This yields four loaves of approximately 2½ pounds each.

Ingredients/Additional Materials:
- ☐ 1½ pounds whole-wheat flour
- ☐ 3½ pounds white flour
- ☐ 1½ Tbsp yeast
- ☐ 1 cup vegetable oil
- ☐ ⅔ cup honey
- ☐ ¼ cup molasses
- ☐ 3 egg yolks
- ☐ 1½ quarts water
- ☐ 2 Tbsp salt
- ☐ Stand mixer with dough hook
- ☐ Rolling pin
- ☐ 3-gallon stockpot or large mixing bowl
- ☐ Four standard loaf pans
- ☐ Plastic wrap

Step 1: Warm the oven to 175 degrees Fahrenheit for fifteen minutes and then turn the oven off but leave the oven light on.

Step 2: Combine the flours and yeast in the bowl of the stand mixer and mix well.

Step 3: With mixer on slow, add the oil, honey, molasses, and egg yolks. Next, add the water slowly and mix just until combined.

Step 4: Turn the mixer off and add the salt.

Step 5: Let the dough sit for twenty minutes before turning the mixer back on—this is the *autolyze* period, which allows the water and flour to blend and helps produce gluten during kneading.

Step 6: Turn the mixer on to medium and knead the dough for seven minutes.

Step 7: Turn out the dough onto a floured surface. Knead by hand until the dough is smooth, usually about four or five times.

Step 8: Place the dough in a well-oiled 3-gallon stockpot or large bowl and then turn the dough over so that the oiled side is on top.

Step 9: Cover the dough with a dishcloth and place it in the warm oven to rise for an hour.

Step 10: Punch down the risen dough and divide into four portions.

Step 11: Grease and flour four loaf pans.

Step 12: Roll out one portion of the dough into a large rectangle, about 12 inches by 18 inches, and fold it into thirds like you would a business letter.

Step 13: Roll the dough out into a rectangle again and then roll it up tightly, beginning at one of the narrow ends.

Step 14: Seal the rolled dough by pinching along the seam, turning it over, and karate-chopping the ends.

Let the dough rise well above the loaf pan.

Step 15: Place the loaf into a prepared loaf pan, cover it with oiled plastic wrap, and return it to the warm oven to rise for about another hour.

Step 16: Repeat Steps 12–15 with the other three portions of dough.

Step 17: When the loaves have risen well above the pan rims but are not slumping down, remove them from the oven and preheat the oven to 350 degrees.

Step 18: When the oven is ready, carefully remove the plastic wrap and slice a shallow groove, centered lengthwise, in the top of each loaf with a sharp serrated knife.

Step 19: Bake the loaves for forty-five minutes on the middle rack.

Step 20: Remove the bread from the pans immediately after taking them out of the oven and let them cool completely—for at least two hours—before slicing, or you'll have gummy bread.

Store whatever you're not using right away in the freezer in plastic bags; the loaves will keep in the freezer for about two weeks. Defrost for about three hours on the countertop before using.

English Muffins

English muffins are very easy to make, and they don't even require an oven. A griddle is good for consistent heating, but they can be done on the stovetop in a pan if you keep a close eye on the heat. Poke the prongs of a fork around the edges of the finished muffins for a "fork-split" texture. This recipe makes ten to twelve English muffins.

Ingredients:
- ☐ ¾ cup warm water (approximately 110 degrees Fahrenheit)
- ☐ 1 Tbsp sugar
- ☐ 2 tsp yeast
- ☐ 3 cups white flour
- ☐ 1 tsp salt
- ☐ 1 egg
- ☐ 1 Tbsp malt vinegar (no substitutes—malt gives the correct flavor)
- ☐ Cornmeal to dust baking pan

Step 1: Dissolve the sugar in the warm water. When the sugar is dissolved, add the yeast and stir to mix.

Step 2: In a large bowl, mix the flour and salt. When the yeast mixture is foamy, add it to the flour mixture along with the egg and vinegar. Mix well.

Step 3: Put the dough on a floured countertop and knead it several times. When it is smooth, return it to an oiled bowl, cover it with a cloth, and set in a warm place to rise for forty-five minutes.

"Dry fry" the cornmeal-coated muffins on the stovetop.

Allow the muffins to cool slightly, and you can serve them warm.

Step 4: When the dough has risen, roll it out on a floured countertop with a rolling pin to about a ½-inch thickness.

Step 5: Cut out circles with a round cutter or tuna can. Reroll the scraps to get as many circles as possible.

Step 6: Put the circles on cornmeal-coated baking sheets. Turn them over to coat both sides with cornmeal.

Step 7: Cover the baking sheet(s) with a cloth and let the dough rise for about another thirty to forty-five minutes, until the circles have doubled in size.

Step 8: Heat an ungreased griddle (or pan) to 325 degrees Fahrenheit.

Step 9: When the griddle is hot, gently place (they can sink if dropped) the muffins on the griddle and "dry fry" them for about fifteen minutes.

Step 10: Gently flip them over (they're less likely to sink now) and dry fry the other side for ten to fifteen minutes or until light brown. Griddle and pan cooking times may vary.

Step 11: Cool the fried muffins on a wire rack.

Serve the English muffins warm with butter or jam—the fork-split texture makes pockets for toppings

to settle into. If you don't eat them right away, you can freeze them in a plastic bag (ours don't usually last!).

Bagels

I got tired of buying expensive four-packs of bagels at the grocery store and realized that I must be able to be make them by hand. They're not quite New York perfection, but they're very tasty fresh from the oven (once they're cool enough!). The price is right, and it's a fun process to learn.

• •

Ingredients:

- ☐ **2 tsp yeast**
- ☐ **3 tsp brown sugar**
- ☐ **1½ cups warm water (110 degrees Fahrenheit)**
- ☐ **4 cups white flour**
- ☐ **1 Tbsp salt**
- ☐ **2 tsp baking soda**
- ☐ **Cornmeal to dust baking pan**

• •

Step 1: Dissolve the sugar in the warm water. When dissolved, add the yeast and stir to mix.

Step 2: Mix the flour and salt. When the yeast mixture is foamy, pour it into the flour mixture.

Step 3: Stir to combine and then knead the dough on a well-floured countertop for five minutes or until it is smooth.

Step 4: Roll the dough into an approximately 12-inch-long tube, lay it on a baking pan, cover it with a cloth, and let it rise for two hours.

Step 5: Place the risen tube on a flour-dusted countertop. With a serrated knife, cut the dough into equal portions of approximately 4 ounces each (I weigh the first one and eyeball the rest). Allow each piece to sit for five minutes.

Step 6: Here's the fun part! Working with each portion, poke your index finger into the middle of the dough piece while it is still on the counter and then spin it around your finger until you create a hole of the desired size.

The bagels will float on top of the boiling water.

Step 7: Set each completed piece on a cornmeal-dusted baking sheet, cover the pieces with a light cloth or a piece of plastic wrap, and refrigerate overnight.

Step 8: The next day, preheat the oven to 450 degrees Fahrenheit. Dust another baking sheet with cornmeal and flour.

Step 9: Fill a large wide-mouthed pot or pan with water, add the baking soda, and bring it to a boil. The pot should be large enough so that the bagels won't touch the bottom or each other when floating in the water.

Step 10: Drop the bagels in small batches into the boiling water—but don't let the bagels touch, or they'll stick together.

Step 11: Boil for two minutes and then turn the bagels over (kitchen tongs are good for this step) to boil on the other side for one minute. They should puff up and feel slightly firm.

Step 12: Remove the bagels with a slotted spoon and place them on a rack until they look dry.

It's hard to wait until the bagels are cool enough to eat!

Step 13: Put the boiled dry bagels on the prepared baking sheet and bake them in the oven for ten minutes.

Step 14: Turn the sheet 180 degrees and bake for another five to ten minutes or until the bagels are golden brown.

Step 15: Remove immediately and let the bagels cool on a rack—they'll be hot!— before cutting them.

For cinnamon-raisin bagels:
- Reduce the salt to 1 teaspoon and add 1 teaspoon cinnamon.
- Before making your dough, bring ¾ cup raisins and ½ cup water to a boil. Remove from the heat, drain, and set the raisins aside to cool. Incorporate the raisins in the initial kneading (Step 3).
- Use flour instead of cornmeal to dust the baking trays.

Grinding Whole Grains into Flour

Store-bought flour is a highly processed food that has been ground, filtered, and often bleached. The best parts of the grain kernel are removed (and sold separately as bran and germ for much higher prices), and less expensive vitamins are added before it is bagged for our use. Why not buy the whole grains, grind them yourself, and get a healthier result from it?

I became a grain-grinding devotee when a German coworker of my husband moved his family nearby. The first time we had dinner at the coworker's house, I noticed containers of whole grains and an interesting kitchen device, which turned out to be an all-in-one unit with various attachments, including a grain grinder. I asked his wife, Bettina, about the grains, and she looked at me as if I were from Mars. She couldn't believe that I didn't grind my own flour, and I couldn't believe that she did.

This project can be done only if you invest in (or borrow) a whole-grain grinder; there's no shortcut. Bettina convinced me that my stand mixer could use yet one more attachment, and I got my own grain grinder. I am

A grinder attachment on a stand mixer.

not yet at Bettina's level of baking, and I still measure obsessively, but I am a convert to grinding whole grains.

When you grind your own flour, grind only as much grain as you'll need for your recipe, as the nutritional quality of the grain begins to deteriorate as soon as it is ground. One cup of whole grains translates roughly to 1¼–1½ cups of flour, depending on the grind you choose. Your flour will be coarser than the refined flour you may be used to, and your baked goods may have a grainy texture (but my chocolate-chip cookies still get eaten!). Freshly ground whole-grain flour lends more of a wheat flavor to the food, which is why it is good in breads and other savory baked goods. Sweet baked goods are usually made with lighter, more refined flours.

Bettina's Advice
- You can grind the grains from rough (coarse) to fine, depending on what you want to do with them. The finest grind is flour, and the coarsest is groats.
- If your grain mill has a grinder made of steel, you can also grind sesame and flax seeds, but only coarsely. Grinding breaks up the seed differently than chewing does, and it makes the vitamins and minerals from the kernel more accessible to your body.
- Many grinders do not grind corn; check before buying.
- For bread, you can use all kinds of grain, but only wheat (and thus spelt, emmer, and other kinds of wheat) contains the gluten that acts like glue in the bread. I prefer wheat for at least half of the flour and whatever I like—oats, barley, rye—for the rest.

Small-Scale Beer Brewing

Making your own beer is incredibly cheap per pint, is really fun, and yields very tasty results. This project requires no special equipment; you'll probably find everything you need already in your kitchen. If you like the process, a whole new world will open up—you'll find a new appreciation for small-batch beers, and you may decide that you want to play with the many different flavors that you can get from home brewing.

Why would you want to do this?
Small-batch, hand-crafted beer has distinct flavors, and it tastes so much better than many store-bought beers. Beer makes a great gift, and brewing can become a fun, rewarding hobby.

Why wouldn't you want to do this?
You don't drink beer, or you don't like beer with interesting or complex flavors.

Is there an easier way?
You can use extracts (instead of whole grains), available at home-brew stores or from online suppliers, but most are formulated for a full 5 gallons. Beer brewed with extracts tastes very different than that made from whole grains, but using extracts can cut out a bunch of steps (similar to using a mix to make pancakes).

Cost comparison:
The price per pint is minimal compared with the price of bottled or pub brews.

Skills needed:
This is pretty easy for the person who knows his or her way around a kitchen. Some knowledge or understanding of brewing and beer qualities can help but is by no means necessary.

Learn more about it:
Whether you were pleased with the results or were slightly dissatisfied but are still curious, by all means, learn more! *The Brewers' Handbook* (Apex Pub, 2000) by Ted Goldammer is a good starting place; also check out the magazines *Zymurgy* (published by the American Homebrewers Association) and *Brew Your Own* (published by Battenkill Communications).

We live in Portland, Oregon, which boasts more brew-pubs per capita than anywhere else in the world and more varieties of small-scale brews than anywhere else in the United States. Given that my husband and I both really enjoy beer and making our own products, it seems almost inevitable that he would eventually take up brewing. He began with extracts—sort of a ready-made "kit" to brew with. Because he's a foodie as well and has a highly developed sense of taste, he moved into whole-grain batches, learning about the subtle nuances that come with having more control over the process and that also yield a more interesting product.

The following is my husband's version of a small-scale batch, which is made without any equipment fancier than what you already have in your kitchen. Ingredients can be bought at a home-brew supply store or from an online supplier. Have your brew shop grind the malt for you; if ordering online, make sure that the supplier grinds the malt and doesn't just ship whole grains.

Experiment with your favorite flavors and styles of beer.

Materials/Ingredients:

- ☐ Large bowl for mash
- ☐ Colander or large strainer
- ☐ Food thermometer
- ☐ Large spoon
- ☐ Two large pots
- ☐ Funnel
- ☐ Two 2-liter soda bottles
- ☐ 1 pound American 2-row pale malt
- ☐ ¾ pound (12 ounces) Maris Otter or British 2-row malt
- ☐ ¼ pound (4 ounces) medium crystal malt
- ☐ 1 ounce Fuggles (hops)
- ☐ 1 packet Muntons brewing yeast
- ☐ Carbonation tablets
- ☐ Water
- ☐ Ice

Step 1: Pour the malt grains into the bowl and mix them together. In a large pot, heat 2 quarts of water to 170 degrees Fahrenheit and pour it into the grain mixture. This is your *mash*. Stir it well, cover it, and let it sit for an hour. During this hour, stir it several times to keep the temperature relatively even throughout. Check the temperature periodically to maintain it at 145–150 degrees Fahrenheit.

Step 2: Near the end of the hour, heat another gallon of water to 175 degrees Fahrenheit and pour a couple of cups into the mash.

Step 3: Place your colander (or large strainer) over the other large pot. Spoon the mash into the colander. Gently pour the water from the mash over the grains, followed by the rest of the gallon of water you just heated. Pour the water over a large spoon so the water splashes over the surface of the grains rather than boring a hole in one spot. You want to rinse as much of the sugar and flavor off of the grains as possible to make a "tea" called *sweet wort*. This wort is the basis of the beer in that it contains the foods that the yeast will convert to alcohol and CO_2 as well as the compounds that form the flavors of your beer. Once you've used all of the water to rinse the grains, you are done with the grains and can use them for muffins, dog biscuits, bird feed, or whatever you like.

Step 4: Bring the sweet wort to a boil on the stove. You may need to add 1 or 2 cups of water to increase the total volume to about 2 gallons at the start of the boil.

Step 5: Once the wort begins to boil, add half an ounce of hops. Boil vigorously, uncovered, for thirty minutes, and then add another quarter ounce of hops. Continue boiling for another twenty-five minutes and then add the rest of the hops. Boil for five more minutes and then remove the wort from the heat. Total boil time is sixty minutes.

Step 6: Follow the directions on the yeast packet to rehydrate the yeast. This is the same process that you use when making bread—adding the dry yeast to a bit of warm water to revive the yeast so it is ready to go to work in your beer.

Step 7: While the yeast is hydrating, place the big pot of hot wort in your sink in an ice bath. Stir the wort to cool it quickly to about 70 degrees Fahrenheit.

Step 8: Pour the cooled wort through your strainer to remove the hops. Don't worry if the liquid has some particles of hops left in it; these will settle out during fermentation.

Boil the wort for thirty minutes after adding the hops.

Step 9: To the cooled and filtered wort, add about half of the hydrated yeast (one packet is enough to make 5 gallons of beer), and stir vigorously to mix the yeast and aerate the wort.

Step 10: Cover the pot with plastic wrap and a layer of aluminum foil and then place the pot in a closet or another dark area with a consistent temperature of 68 degrees Fahrenheit. As the yeast goes to work, it creates CO_2 that will push its way out of the plastic wrap; this is desirable. The wrap is mainly to keep air and airborne impurities out. The pressure inside the pot will let gas out and keep the baddies from getting in, and the foil assists in keeping the plastic wrap in place.

Step 11: Wait three to five days for the yeast to be done. You'll see no more bubbles on the surface of the wort (now beer!), and the yeast will be settling down to the bottom of the pot. I like to brew on a Sunday so I can move to the next step on the following Saturday; this always allows enough time for fermentation to complete and the yeast to settle.

Step 12: Sanitize the bottles with a diluted bleach solution of ¼ to ½ capful of bleach to a full 2-liter bottle of water. Let the solution sit in the bottles for ten minutes and then empty and rinse them with hot water. Pour one of the bottles out through your funnel to sanitize it, too.

Step 13: Add three or four carbonation tablets to each sanitized bottle.

Step 14: Fill the bottles with your beer by ladling it into the funnel. Try not to disturb the layer of yeast that has accumulated on the bottom of your fermentation pot.

Step 15: Seal the bottles tightly so that the CO_2 created as the remaining yeast in the beer consumes the carbonation tablets stays in the bottle and carbonates the beer. Put the bottles back in the area where you fermented and let them carbonate for a week or so.

Step 16: Chill and drink. The beer may not be as bubbly as store-bought beer, but it will resemble a real English ale and should be tasty!

Fruit Cordials

I love summer berries, but I'm not a big jam or preserves eater. I like to preserve the look, color, and taste of these summer treats in a different way. Berry cordials make wonderful gifts and, with their beautiful colors, are very pretty to have on display. There is something special about presenting a gift of a liqueur that you made yourself, and sipping on a cordial is a nice way to end a meal.

Why would you want to do this?

This is an easier way to preserve berries than making jams or preserves, and the cordials can be used in many ways, not just in drinks. They can be given as gifts, used as natural cough syrups, or poured over ice cream or cake as luxurious dessert toppings. Some flavors can also be used as ingredients in exotic marinades or glazes.

Why wouldn't you want to do this?

You don't like sweet drinks, you don't drink alcohol, or your family and friends don't drink alcohol.

Is there an easier way?

There's not much about this that can be easier. Straining out all of the residue is the biggest challenge, and that just takes time and patience.

Cost comparison:

Your homemade version will cost about half of what a store-bought bottle of liqueur costs—even less if you use your own homegrown fruit.

Skills needed:

No special skills needed—it's pretty darn easy.

Further refinements/learn more about it:

You can take this to a higher level by experimenting with different fruit flavors or trying recipes for flavors other than fruit, such as Irish cream or coffee. Two good references are *Making Liqueurs for Gifts* (A Storey Country Wisdom Bulletin, 1988) by Mimi Freid and *The Joy of Home Wine Making* (William Morrow, 1996) by Terry A. Garey.

The book that I use as my cordial/liqueur bible, *Making Liqueurs for Gifts* by Mimi Freid, distills (pun intended) the cordial-making process down into three simple steps: steep, strain, and filter. That's about it.

The final yield of this formula will be a little more than a pint of cordial (you won't be using the entire quart of vodka because volume will be taken by the berries). These proportions can be easily adjusted, depending on what you have available.

When you collect your materials, try to use fruit that is ripe but not overripe. Either pick your homegrown fruit or get fresh fruit from a pick-your-own farm, a farmers' market, or a grocery store while the fruit is in season. "Off" flavors can be transmitted to the drink easily, and you want your cordial to have the best fruit flavor possible.

Materials/Ingredients:

- ☐ Quart jar with lid
- ☐ Strainer or fine sieve
- ☐ Funnel
- ☐ Paper towels or coffee filters
- ☐ Decorative, resealable bottles
- ☐ 1 cup sugar
- ☐ 1 quart berries
- ☐ 1 quart vodka

Step 1: Rinse the fruit immediately prior to use, no earlier. Moisture can cause mold to form on the fruit, and although high-proof alcohol has antibacterial properties, the moldy flavor can be imparted to the liqueur.

Step 2: Pour the sugar into a quart jar. Fill the remainder of the jar with the fruit, being careful not to pack it down or squish it. Fill the jar with vodka so that the fruit is completely covered and then seal the jar with a lid. I gently roll the jar around a bit to mix the sugar and make sure it is evenly moistened. It will settle to the bottom again until it fully dissolves.

Step 3: Let the jar sit for two months in a cool, dark spot, such as a basement or pantry closet, gently shaking or turning it about once a week to mix the

ingredients. The sugar will slowly dissolve, and the berries will release some of their color into the liquid.

Step 4: At the end of the two months, strain the fruit in a mesh strainer or fine sieve, reserving the liquid. You can serve the fruit in small quantities over ice cream or pound cake, or you can squeeze or press it to get more of the liquid out (be forewarned: if you do this, there will be more residue to filter out).

Step 5: Pour the reserved liquid through a finer filter—coffee filters or a paper towel in a funnel will work. This step takes a long time, and you may have to change the filter once or twice when it stops letting the liquid through. If you skip this step, you will see residue form on the bottom when the cordial is bottled. If you'd prefer, you can decant the clear uppermost liquid and pour just the sludgy parts through a filter. A crystal-clear, beautiful cordial is the goal.

Store in bottles with corks or stoppers. Because the cordials take on the beautiful fruit colors, you can display them in decorative bottles. To give one as a gift, make a nice label or wrap raffia around the neck of the bottle and attach a tag. Depending on the bottle, you might even be able to dip the closed top in wax for an attractive seal similar to that on a fine Madeira or port.

Homemade Mixers

If you look at the ingredients in ready-made cocktail mixers or the prices of the more wholesome versions, you'll be happy to try making your own. Once you taste them, you'll never go back!

Why would you want to do this?

Have you looked at the ingredients in premade mixers? I don't want that stuff in my drinks! Plus, you'll save money and impress your friends with your delicious homemade cocktails.

Why wouldn't you want to do this?

You don't drink or serve mixed drinks, or you like the taste of store-bought mixers.

Is there an easier way?

The only easier way is to simply buy premade mixers, but in this case, the enjoyment and taste of homemade are too good to miss.

Cost comparison:

The cost per ounce is much lower. For example, for about what it costs to buy a 1-liter bottle of margarita mix, which makes six drinks, you can make a batch of your own that makes approximately twenty-four drinks.

Skills needed:

Basic cooking skills and patience.

Having a beautiful, well-stocked bar at a party is so much fun (as long as you have friends who enjoy themselves responsibly), and for people who enjoy food and drink, the flavors of homemade mixers are great additions. You don't have to drink alcohol to enjoy the full, bright flavors of ginger and lime without additives such as high-fructose corn syrup. Once you've tasted homemade mixers, it's darn near impossible to enjoy a store-bought version.

Materials for Each Project:
- [ ] Pots
- [ ] Saucepans
- [ ] Stirring utensils
- [ ] Colander or strainer
- [ ] Coffee filters or reusable fine coffee filter basket
- [ ] Funnels
- [ ] Bottles for storing finished product

Tonic Concentrate

Ingredients:
- [ ] 4 cups water
- [ ] 3 cups cane sugar
- [ ] 3 Tbsp cinchona powder (the source of quinine; look for it online)
- [ ] 6 Tbsp food-grade citric acid (available at natural-food stores or where cheese-making supplies are sold)
- [ ] 2 limes, zested and juiced
- [ ] 2 stalks (leaves and stem) lemongrass, chopped coarsely

Step 1: In a saucepan, bring the sugar and water to a boil until the sugar dissolves. Reduce heat.

Step 2: Add the cinchona powder, citric acid, lemongrass, lime zest, and lime juice. Stir gently and then simmer for twenty-five to thirty minutes, until the powder appears to dissolve and the syrup is thin and runny.

Step 3: Remove from heat and allow the mixture to cool a bit. Strain the large pieces out through a colander or a coarse strainer.

Step 4: This is the time-consuming part. Pour the syrup through a fine filter to remove the remaining cinchona dust. Since the powder is very fine, this takes a long time, and the sticky syrup can be very messy if it spills. To speed up the process, filter the syrup initially through several overlapping layers of moistened and wrung-out cheesecloth, rinse the cheesecloth, and do it again. The more layers of cheesecloth, the more powder will be filtered out. For the final filtering, pour the syrup into a moist coffee filter (preferably supported in a single-cup-type filter contraption or a funnel) until all of the debris has been filtered out. You may have to filter the syrup several times and change filters once or twice; remember to wet the filter before filling it with the syrup.

Store the concentrate in a resealable bottle in the refrigerator (I use a glass quart milk bottle). I'm not sure of the exact shelf life; we always use ours up before it goes bad!

To use, pour one ounce in a glass, top with ice cubes, add gin or vodka (optional), and fill the remainder of the glass with carbonated water. Stir well, garnish with a lime wedge, and enjoy!

Ginger Ale Concentrate

Ingredients:
- [ ] 6 cups water
- [ ] 3 cups cane sugar
- [ ] 2 cups (4 medium to large roots) fresh ginger, peeled and chopped coarsely
- [ ] Small hot pepper, finely chopped (optional)

Step 1: Mix all ingredients in a saucepan and bring to a boil. Reduce heat and simmer for an hour or so. The syrup will reduce somewhat—that's OK.

Step 2: Remove the pan from the heat, and let the mixture cool. Pour it through several layers of cheesecloth to remove the chunks.

The fresh ginger in your homemade ginger ale concentrate is a soothing digestive aid.

Store in a bottle in the refrigerator. To use, pour 2 to 3 tablespoons of syrup (to taste) into a glass, top with ice cubes, and fill the remainder of the glass with carbonated water. Stir well and garnish with a lemon or lime wedge or a chunk of candied ginger. This drink is very good for digestion or stomach upsets (minus the hot pepper and citrus garnish) and can boost the immune system.

Margarita Mix

I know that many people make margaritas without a mix, but if you are hosting a party and plan to make batches of margaritas, it's good to have the basics of the drink taken care of. You can then just pour the mix into the blender or over ice in glasses and add tequila and triple sec, and you're done.

• •

Ingredients:
- ☐ 3 cups water
- ☐ 3 cups cane sugar
- ☐ 1½ cups fresh lemon juice
- ☐ 1½ cups fresh lime juice

• •

Step 1: In large saucepan, bring the water and sugar just to a boil, until the sugar dissolves.

Step 2: Remove the pan from the heat. Allow the syrup to cool and then mix in the citrus juices.

Store in a bottle in the refrigerator and use within one week. For frozen margaritas, fill a blender with ice and then pour in chilled syrup about halfway to two-thirds of the way up. Add liquor to preference and blend until smooth. Serve in glasses with salted rims. For a margarita on the rocks, pour approximately ¼ cup of chilled syrup over ice. Add liquor to preference, and garnish with a lime wedge.

Your margarita mix will have a sweet, fresh, citrusy flavor.

Make Your Own Tea

One of the best uses for homegrown herbs is in tea, and fresh herbal tea is one of the best kinds of tea.

Why would you want to do this?
You can achieve an unlimited variety of blends, your homemade tea tastes better than store-bought tea, and making your own tea bags is an easy, quick project as well as a fun gift idea.

Why wouldn't you want to do this?
You don't drink tea, or you don't want to spend time making your own.

Is there an easier way?
If you make a big batch of tea bags at one time, you'll be set for a while. Another option is to crumble your favorite herbs or herb mixes and keep them in a dark, airtight container in your kitchen. When you want a quick cup, you can scoop out the leaves on the spot.

Cost comparison:
You can make 100 tea bags for about the same price as a store-bought package of 20 tea bags; the cost is even less if you grow your own herbs.

Learn more about it:
The Herb Tea Book (Interweave Press, 1998) by Susan Clotfelter is a great compilation of recipes and ideas for making and using your own teas and tea blends.

This type of tea infuser sits at the bottom of the cup or mug.

My friend Bridget makes tea with her own herbs. I told her once, at one of our many teatimes together, how much I admired the practice, and she looked at me, completely flabbergasted. When I sheepishly confessed that I still bought tea, she exclaimed, "I can't believe that anyone who has herbs doesn't make [his or her] own—it's ridiculous not to!"

That moment was my moment of change. I can understand buying tea from the same perspective as I can understand buying any of the things that I talk about making in this book. It's faster to grab a box of tea than it is to produce it yourself, especially if you want a variety of flavors. But really, once you've made your own tea and have seen how easy it is, how much money you can save, and how much better the tea tastes, you'll understand how Bridget felt.

The first step to making tea is drying your herbs, which I discuss in Section V, Project 1. You can brew tea with dried herbs by using infusers, strainers, or bags.

Infusers sink down into a cup or pot and come in many forms; for example, there are spring-controlled

Common Herbal Tea Ingredients

unless specified that they should be dried, herbs can be brewed either fresh or dried

CAMELLIA (*Camellia sinensis*)
PARTS USED: leaves, dried properly
This is grown in India and China for the well-known English Breakfast and Darjeeling varieties of caffeinated tea that many of us are familiar with. Green tea and black tea also come from this plant but are dried to different degrees. *C. sinensis* is an easy shrub to grow and keep small, and it contains many powerful antioxidants. It is the only ingredient listed here that is a caffeinated addition to tea.

MINT (*Mentha* spp.)
PARTS USED: leaves
Any species of mint, including spearmint and peppermint, is great for making tea. The taste and aroma of mint leaves can invigorate and energize, and they are beneficial for digestion.

LEMON BALM (*Melissa officinalis*)
PARTS USED: leaves
Melissa looks and behaves like mint and is a good companion for mint tea blends, but it has a strong, almost sweet, lemony flavor. It is a digestive herb, and it also serves to strengthen the immune system.

GERMAN CHAMOMILE
(*Matricaria recutita*)
PARTS USED: flowers
Chamomile is a traditional bedtime tea because its herbal qualities invoke feelings of calm and relaxation. It has a sort of dusty, tangy flavor that appeals to many, and it is often used as a digestive aid.

CATNIP (*Nepeta cataria*)
PARTS USED: leaves, flowers
Although the catnip plant is in the mint family and is used to make cats playful and energetic, it has the opposite effect on humans, serving as a powerful soother in tea. A blend of tea with catnip helped me through a particularly stressful home-remodeling project.

A tea strainer sits across the top of the teacup.

mesh spoons as well as infusers with screw-on lids and chains to hang from the lip of your cup or pot. Silicone infusers, which are growing in popularity, come in a lot of fun shapes. Infusers are nice in that there is no waste when the tea is done steeping, and the herbal matter itself can go straight into the compost. (Tea bags can be

Another type of infuser hangs from a chain.

Cinnamon

good flavor. Crushed whole cloves are stronger and can impart a bitterness to the finished product. I sometimes put whole cloves into the blend when making a pot for more than four cups, but I use only three to five cloves. Clove and cinnamon are both used in mulling spices, and they get me into the spirit of the season around the winter holidays.

CINNAMON (*Cinnamomum* spp.)
PARTS USED: sticks
Cinnamon sticks are one of the few tea ingredients that you'll have to buy rather than grow yourself. You can use cinnamon sticks to stir your tea, or you can crumble them into a tea blend to be steeped. Powdered cinnamon is too fine and

is very strong; it will muck up your concoction and contribute a very bitter taste.

CLOVES (*Syzygium aromaticum*)
PARTS USED: dried flower buds
As with cinnamon, you'll have to purchase cloves for your tea and use them carefully to achieve a

ECHINACEA (*Echinacea purpurea*)
PARTS USED: flowers, leaves
Echinacea is an easy-to-grow perennial plant that has gained fame for its curative properties; I swear by echinacea tea during cold and flu season. Like chamomile, it acts as a soothing ingredient and can have a similar dustiness in flavor, but it has a slightly more floral overtone.
CONTINUED ON PAGE 58

composted also, but you have to pay attention to what they are made of. Silk tea bags are gorgeous, but they are expensive and do not compost.)

A tea strainer sits on top of your cup or pot and holds the loose tea without a lid; you pour your boiled water straight through it. A strainer is easier to clean than an infuser is, but depending on the depth of the strainer, it may not hold the leaves in the liquid and will thus yield a weaker decoction.

You can buy empty tea bags to fill—some natural-food stores have them in the tea aisle near the loose tea sold in bulk, or you can order them online. Some bags have a simple double-fold flap that you fold over the top after filling the bag with herbs; others can be stapled closed after folding. My favorite is the type that seals shut with a quick pass of an iron across the open end. A box of homemade tea bags makes a unique holiday or hostess gift.

The following instructions are for brewing a cup of tea using your own dried herbs. Keep in mind that homegrown and dried herbs often have stronger flavors than store-bought herbs because they are fresher and most likely dried in gentler conditions, so you may choose to adjust your quantities or steeping time accordingly. The chart on pages 56–59 explains the names and properties of many herbs commonly used in tea, but it is by no means exhaustive. There's much more to learn about herbs, their myriad uses in teas, and their many other benefits.

Materials/Ingredients:
- ☐ **Tea infuser, tea strainer, or fillable tea bags**
- ☐ **Dried herbs of your choice**

Step 1: Fill the infuser, strainer, or tea bag with 1–2 tablespoons of the desired herb or variety of leaves and place on or in the mug.

Step 2: Pour freshly boiled water through the strainer or over the bag or infuser in the mug.

FENNEL (*Foeniculum vulgare*)
PARTS USED: seeds
In India, people eat fennel seeds after meals as a digestive aid. When added to tea blends, the seeds impart gentle licorice overtones that I find subtly invigorating.

LEMON VERBENA (*Aloysia triphylla*)
PARTS USED: leaves
This herb is tricky to grow in colder climates, but if you can grow it, you'll never regret it. Lemon verbena is wonderful to cook with, and it imparts a more pungent, direct lemon flavor in teas than lemon balm does. It is a calming herb that aids digestion and sleep.

LEMON PEEL
PARTS USED: peel (also called *zest*)
A squeeze of lemon juice in brewed tea is hardly a novel idea, but you can also use lemon peel, fresh or dried, in your tea blend. Go light on the peel, as the essential oils in it can overpower the mix easily. Orange peel can be used the same way.

ROSE, WILD (*Rosa* spp.)
PARTS USED: leaves, rosehips (seedpods)
Living in the City of Roses (Portland, Oregon), I learned early on that the lovely hybrid tea roses that are used in bouquets and displays are not for candying, steeping, or adding to concoctions; wild roses are the roses that we can ingest. Rose petals in tea give a lovely, gentle floral scent and can serve as a digestive aid. Rosehips are a powerful source of vitamin C and useful with echinacea in tea for a cold or flu.

Rosemary

Step 3: Let the herbs steep for three to five minutes, depending on your desired strength. Stronger herbs shouldn't steep for much longer than five minutes, as they will impart a bitter taste.

Step 4: Remove the strainer, infuser, or bag. Sweeten and add milk to your tea as desired. **Note:** The herb stevia (see chart below) is a natural sweetener, and you can include it right in your tea blend.

ROSEMARY (*Rosimarinus officinalis*)
PARTS USED: (needle-like) leaves
Generally more of a savory herb, rosemary can be judiciously added to a tea blend to impart its strong, invigorating scent. Rosemary acts as an herbal antidepressant, digestive aid, and mild stimulant.

SAGE pineapple (*Salvia elegans*) and clary (*Salvia sclarea*)
PARTS USED: flowers, leaves
While the extensive sage family is used for both culinary and medicinal purposes, these two varieties have sweeter essences and are better additives to tea than some of their cousins.

SCENTED GERANIUMS
(*Pelargonium* spp)
PARTS USED: flowers, leaves

These are not your grandmother's bright-red Martha Washington geraniums. Scented geraniums are a different family, and they have a wide variety of scents and flavors as the result of extensive hybridization. They, like roses, can impart a sweet, gentle floral overtone to your tea.

STEVIA (*Stevia rebaudiana*)
PARTS USED: leaves, flowers
Stevia is a natural sweetener, having many times the sweetness of refined sugar; it also aids in digestion and has many other uses. It is challenging to grow in colder climates but can be a houseplant.

THYME common (*Thymus vulgaris*) or lemon or lime thyme
(*Thymus citriodorus*)
PARTS USED: leaves and flowers

Lemon or lime thyme is a better addition to tea than the garden-variety culinary thyme, but any type of this herb will contribute a pungent flavor to the tea. Thyme is a familiar and potent healing herb that is used to help ease the symptoms of colds and coughs through ingestion.

NETTLE (*Urtica dioica*)
PARTS USED: leaves
Use caution when harvesting fresh nettles because small hairs on the plants produce a stinging reaction. They do not sting when they are dried, and they are strong herbs for respiratory and circulatory health, possessing anti-inflammatory qualities. They are rich in vitamins and minerals—and also make a great pesto!

Canning and Pickling

Buying packaged nonperishable food from the store is not the only way to have food on hand outside the refrigerator. You can preserve your own creations easily and cheaply.

Why would you want to do this?

Canning and pickling allow you more ways to enjoy your homegrown/homemade foods and prolong the enjoyment of your harvest (that you've saved money on by growing yourself). Home-preserved foods are healthier than store-bought nonperishables because you are using a preservative *method*, not preservative *ingredients*.

Why wouldn't you want to do this?

You don't want to store the canning jars, or you don't have space to store food long-term.

How does this differ from the store-bought version?

Home-preserved food tastes homemade, and you get the pride of having done it yourself.

Is there an easier way?

You can buy home-preserved foods from hobby growers or small farms for the nutrition and taste of homegrown food but not the cost savings.

Cost comparison:

There is the upfront cost of acquiring the jars (and kettle or pressure cooker if you need one). Once you have the materials, it is virtually free.

Skills needed:

Basic cooking skills.

Learn more about it:

Can It! (I-5 Press, 2012) by Jackie Callahan Parente; *Well-Preserved* (Clarkson Potter, 2009) by Eugenia Bone; *The Joy of Pickling* (Harvard Common Press, 2009), revised, by Linda Ziedrich; *The Beginner's Guide to Preserving* (free PDF at www.homestead harvest.com) by Dena Harris and Nicole Taylor.

Preserving food from your own garden in your own kitchen is a rewarding project that goes hand in hand with growing and cooking the food. The best thing is enjoying the fresh harvest with appreciation for what went into it. The second-best thing is appreciating it again when the warm days and the toils of the garden are forgotten and many of the flavors of harvest season are gone for the year. You can enjoy your healthy bounty again in the winter and early spring, reducing your grocery budget and extending the enjoyment of your homegrown food.

If you don't have a garden, you can go to a pick-your-own farm or a farmers' market and stock up during the peak of the season. Even retail operations often sell bulk quantities of produce at reduced prices. Take advantage of the savings and preserve the excess.

Canning

Pressure canning, or heat preserving, is a method of food preservation that long predates refrigeration. By heating containers and their contents to a threshold beyond the tolerance of bad bacteria and sealing them to prevent the entry of pathogens, food can be preserved for years in room-temperature conditions. Most of the flavors and nutrients are retained in preserved foods.

Do some research on preserving your foods of choice before getting started. Certain foods need higher heat or acidifying additives to eliminate the possibility of bacteria growth. Eliminating bacteria growth is critical to preserving the food; otherwise, you'll end up with spoiled food and maybe even exploding jars!

When a jar is properly canned, the lid is tightly sealed to the jar without the outer ring attached.

Glass jars explode when microbes growing in a sealed container create a gassy buildup of pressure that eventually causes the jar to burst. One way to sidestep the possibility of explosion (besides doing everything you can to preserve the food correctly) is to remove the canning rings when you seal the jars. If properly sealed, the lids will stay on without the threaded rings. Without the ring, any pressure inside the jar will cause the sealed lid to simply pop up or off. If I ever find a jar with either a loose lid or mold, I know not to use it, and I've avoided the mess of shattered glassware.

The food that most people (us included) seem to mourn at season's end is tomatoes. Let's face it, the tomatoes available for purchase in grocery stores just don't hold a candle to what comes out of our gardens or from the farmers' markets. However, while it won't beat a tomato served fresh from the garden, fresh-preserved tomatoes are one of the best ways to carry that summer flavor through the year. Canning your own tomatoes is easy. We make tomato sauce and sometimes whole stewed tomatoes (when we are up to our eyeballs in the harvest) during the flush of summer, often canning as many as we eat. This project focuses on tomato sauce.

For years, we used pint jars for tomato sauce because it was easy to fit five into our pressure cooker. We would use two per spaghetti or pizza meal, and I still find myself grabbing two jars even though we've now moved up to quart jars. Most urban farmers are best suited to start with pints—at 16 ounces, they mimic the 14.5-ounce

Canning jars have a two-part lid with an outer ring.

cans of store-bought sauce—and they are easy to store both filled and empty.

The following method of canning is the *boiling-water-bath method*. As the food in the jar is heated to boiling by the surrounding water, the air is pushed out, creating an environment of relative sterility. When the jars are removed from the bath and cooled, the lids are vacuum-sealed onto the jars, which is why the rings can be removed for storage.

Tomato Sauce

• •
Materials/Ingredients:
- ☐ Glass canning jars with two-part lids (lid plus ring), 8 jars per gallon of sauce
- ☐ Large kettle
- ☐ Pots/pans
- ☐ Clean cloths
- ☐ Optional: Funnel
- ☐ Optional: Pressure canner
- ☐ Homegrown tomatoes (as many as you want to make into sauce)

• •

Step 1: Wash your jars well. This step can be ideally completed by timing the jars to come out of the dishwasher just as you are ready to ladle the sauce into them so that they'll be clean and already hot (cold glass can crack or break when receiving hot contents).

Step 2: Cook your tomato sauce as you normally would (see the sidebar if you don't already have a recipe), omitting meat products (it is riskier and more detailed to can foods with meat in them, so you can tackle that project when you're more advanced). You can put in herbs and seasonings or just tomatoes, depending on how you'll use the sauce. Because we use ours in chili, as spaghetti sauce, as pizza sauce, and in salsa, I use plain tomatoes and add the seasonings after I open the jars.

Step 3: Put the flat lids for the jars into a small pan of boiling water to sterilize them and soften the rubber rims. Start a big kettle (large enough to hold at least one or two jars fully submerged under bubbling water) of water boiling on the stove.

Step 4: Fill each of the hot, clean jars with sauce no higher than where the "shoulders" of the jar become the vertical mouth edges (this air gap is called the headspace). If you have a wide-mouth funnel, use it; otherwise, any spills on the rim or threads of the jar should be wiped off with a clean, damp cloth before lidding.

Step 5: Place a hot lid on top of each cleaned-off jar and screw on the rings, tightening to "finger tight" (a term that always warrants discussion—don't wrench it tightly, just screw it on as firmly as you would a regular jar lid).

Step 6: When all of the jars are full, place as many as you can into the kettle of boiling water, making sure that

Homemade Tomato Sauce

When you are facing 30 or 40 pounds of tomatoes, it helps to make a quick batch of sauce before the fruit flies find the tomatoes. Here's my tried-and-true method, most easily done with a food mill:

Cut each tomato into chunks, removing the stem ends. Put them all in a big pot on a medium burner. If there isn't much juice, add a bit of water to keep them from scalding on the bottom. When the tomatoes begin to heat up (you'll see steam coming from the pan), stir and mash them. Keep cooking them until they lose their shape and are bubbling.

Set up the food mill over a large bowl. Ladle the cooked tomatoes into the mill in batches, milling out the skin and seeds. The sauce comes out of the mill and into the bowl.

If you don't have a food mill, drop each whole tomato into boiling water for ten seconds or so. Remove each tomato with a slotted spoon and peel off the skin (it should come off easily; if not, put the tomato back into the boiling water for a few more seconds). Cut the peeled tomatoes into quarters, removing the stems and seeds. Put the pieces in a large pot and cook down into sauce.

Okra pickled with mustard seeds and spices.

the boiling water covers each jar fully. Keep the water at a boil for twenty minutes (boiling time varies depending on the type of food you're preserving).

Step 7: Remove the jars from the bath and let them sit on the counter until cool. When properly sealed, the lids should be sucked down tightly onto the jars. You might hear the telltale "ping" of the suction as the jars cool. Push gently on each lid before you remove the ring to make sure it doesn't bounce—a properly sealed jar should have a taut, inverted lid.

Step 8: Reprocess each improperly sealed jar, if any, by removing and wiping off the lid, ring, and jar rim; replacing the lid and ring; and boiling the jar for another twenty minutes. Alternatively, you can refrigerate these jars and use their contents within a week.

Step 9: Remove the rings from the properly sealed jars and label them with the month, day, and year. Store them in a cool, dark environment for up to a year (enough time for the next tomato harvest!).

Pickling

Pickling is a different animal than canning because it changes the food's flavor and thus isn't suited to all foods or all palates. Pickling preserves foods by creating an environment hostile to bad bacteria through the high acidity of vinegar and the saline content. In some formulations, the fermentation produced by putting edibles into this environment also generates beneficial bacteria that aid digestion. Devotees of pickling and fermenting insist that these methods offer myriad health benefits, and they are popular among certain circles, such as paleo-diet and raw-food practitioners.

We love sauerkraut, and pickled beans are tastier than their frozen counterparts, retaining a great snappy crunch. "Dilly beans" are one of our favorite foods, and I often find that my kids have opened a recently prepared jar while I'm still picking plenty of fresh green beans from the garden. Frozen beans can get soggy when cooked, and this recipe is a great way to preserve the bounty's crispness.

Dilly Beans

Materials/Ingredients:

- ☐ **Large pot**
- ☐ **Eight 1-pint canning jars/lids**
- ☐ **Optional: Pressure canner**
- ☐ **3 cups white vinegar**
- ☐ **3 cups water**
- ☐ **½ cup pickling salt**
- ☐ **3 pounds green beans**
- ☐ **Approximately 8 generous sprigs of dill leaves (unopened flower heads are pretty also)**
- ☐ **50–60 whole peppercorns**

Step 1: Bring the water, salt, and vinegar to a boil, and then remove from heat. Set aside.

Step 2: Remove the strings and stem ends from the beans, breaking or cutting them into similar lengths that will fit in the pint jars. Pack the beans into the jars so that the beans are standing up inside the jars. Place at least one sprig of dill in the midst of the standing beans.

Step 3: Drop five to eight peppercorns into each jar.

Step 4: Ladle the salt/vinegar mixture over the beans, covering the beans entirely and filling the jars to their shoulders. Put the lids and rings on the jars.

For a variation on this recipe, you can make basil beans by putting in sprigs of basil and a clove or two of raw garlic instead of dill and peppercorns before adding the salt/vinegar brine.

Properly pickled foods should not necessitate a boiling-water bath or pressure canning, but I like to use one of these methods as "back-up" preservation if I will be storing the jars at room temperature. You can do the boiling-water bath as described in the foregoing tomato-sauce project. I process the jars in a pressure canner for ten to fifteen minutes to create the vacuum seal.

Sauerkraut

• •

Ingredients/Materials:
- ☐ **Large (3- to 5-gallon) food-grade tub or vat**
- ☐ **Large bowl**
- ☐ **Clean cloth**
- ☐ **Optional: Pressure canner and canning jars/lids**
- ☐ **1 or 2 large cabbages (enough to make 5 pounds)**
- ☐ **Up to 3 Tbsp pickling salt**

• •

Step 1: Shred the cabbage as finely as possible (or to the coarseness that you like for sauerkraut). Put it in a large bowl and sprinkle the salt over it. Combine and mix the salt and cabbage together well.

Step 2: Put the cabbage/salt mixture into the tub. Within a half hour or so, the salt should cause the cabbage to release liquid (fresh cabbage has more natural liquid; if the cabbage has not been recently harvested, it will be drier). If you find that you need more liquid, add a brine of two parts pickling salt dissolved in three parts water.

Step 3: Pack the mixture down tightly in the tub, making sure that it is covered entirely by liquid. You'll need to put something on top of the cabbage to keep it fully immersed in the brine—a plate with a slightly smaller diameter than the tub works well and can be held down with a clean heavy can or bottle on top. The purpose is to keep air away from the cabbage so that it can ferment.

Step 4: Cover the container with a clean, thin cloth and let it rest in a cool, dark place. Check the sauerkraut daily, making sure that all of the cabbage remains submerged. Add more brine if needed. If a scum forms, skim it off with a spoon, wash the plate and whatever you have weighing it down, and replace.

Step 5: Wait two to four weeks for fermentation to be complete.

Store the finished sauerkraut, tightly covered, in the refrigerator or a very cool place (40 degrees Fahrenheit maximum). For room-temperature storage, can the sauerkraut in jars as described in the tomato-sauce project, processing pints for twenty minutes.

You can sanitize and reuse glass jars for your pickled sauerkraut as long as the lids close tightly.

Drying Fruit

Store-bought dried fruit can be rather expensive and quite often contains sugar and/or sulfur (to prevent browning), which have undesirable side effects on our digestive systems. Although using a commercial dryer really is the easiest option, you can dry fruit in your own oven.

Why would you want to do this?
Drying allows you to preserve the bounty of a harvest or a bulk produce purchase and prolong the taste of the season, and dried fruit does not take up refrigerator or freezer space.

Why wouldn't you want to do this?
You don't like dried food or you don't want to try drying in your oven or buy/borrow a dryer.

How is this different from the store-bought version?
Packaged dried fruit is often not entirely dry, so your homemade version may be drier and a little less flavorful than what you can buy. Manufacturers keep the fruit soft by vacuum-packing and/or using preservatives, but isn't that what you're trying to avoid?

Is there an easier way?
If you are serious about wanting to dry fruit, consider buying a food dryer or sharing the purchase of one with a friend. A food dryer is designed specifically for drying food, and it uses much less energy than the oven does.

Skills needed:
Basic cooking skills.

Learn more about it:
Food Drying Techniques (A Storey Country Wisdom Bulletin, 1999) by Carol W. Costenbader; *The Big Book of Preserving the Harvest* (Storey, 2002) by Carol W. Costenbader.

Almost everyone eats dried fruit, sometimes without even realizing it; for example, those raisins in your breakfast cereal are dried grapes, but they are so common that they almost seem like a fruit unto themselves. Dried fruit is ubiquitous in packaged trail mixes and has become much more common in salads.

Drying is a great way to preserve food, especially fruit, because the drying process concentrates (not increases) the sugars in fresh fruit, resulting in a sweeter taste. This project describes how to try oven drying before you commit to the purchase of a food dryer.

Home-dried fruit makes a healthy, preservative-free sweet snack.

Dried Apples

• •

Materials/Ingredients:
☐ Cooling racks
☐ Plastic storage containers or freezer bags
☐ Apples (less juicy varieties, such as Golden Delicious, Jonathan, and Gala, will dry more easily)

• •

Step 1: Seed, core, and cut apples crosswise into slices of uniform thickness, about ¼–½-inch thick.

Step 2: Lay out the apple slices on the cooling racks, making sure that the slices don't touch each other. Cookie trays will work, too, if you flip the slices over periodically during drying to expose the other sides of the apples.

Step 3: Heat the oven to 150 degrees Fahrenheit (or your oven's lowest setting if it's higher than 150 degrees). Put the cooling racks with the apple slices into the oven, centering them on all available oven racks and allowing space all around them for air circulation.

Step 4: Check the apples regularly, rotating the trays and switching oven racks (and turning the slices, if needed) to obtain even heating.

Step 5: When the slices are dry (after about 10 to 20 hours), remove them from the oven. If you'll be storing them at room temperature, condition them (see next paragraph); otherwise, freeze them in plastic bags or other suitable containers once they've cooled completely.

To condition your dried apples, seal them in a plastic container once they are completely cool. Open the container and stir them once a day for seven to ten days to help distribute any remaining moisture. If moisture appears inside the container or lid, check that there is no mold in the container and let the slices air-dry longer.

Freezing Produce

If you don't want to can, pickle, or dry your fresh produce, the easiest way to preserve it is to toss it in the freezer. Anyone can do this quickly and successfully.

Why would I want to do this?

This is the quickest and easiest method of preserving an abundance of produce. Produce that is frozen soon after being harvested is second only to fresh in nutrient quality.

Why wouldn't I want to do this?

You don't eat at home often, or you don't have any freezer space.

Cost comparison:

If you've grown your own vegetables and fruits, freezing them extends the time in which you can enjoy them, thus saving you the costs of buying produce.

Difficulty level:

Freezing most foods is very simple. In fact, depending on the food, it can be a good chore for children.

Learn more about it:

The Big Book of Preserving the Harvest (Storey, 2002) by Carol W. Costenbader.

Produce that is transferred from field to freezer rapidly and frozen as quickly as possible retains much of its nutrient content. Freezing is incredibly easy and economical, and it generates less waste than buying packaged frozen food does. Almost any fruit or vegetable can be frozen. There are different ways to prepare the produce before freezing and different ways to use the frozen produce once it is thawed.

Materials/Ingredients:

☐ Freezer
☐ Cold-stable food containers and/or freezer bags
☐ Kitchen utensils
☐ Cookware
☐ Homegrown produce

Fruit or Vegetable	Preparation before Freezing	Use after Defrosting
Apples: slices	Peel, core, slice. Freeze slices on tray, then place in bags. May brown a bit in freezer.	In pies, tarts, cobblers, smoothies; placed on ham while baking
Asparagus	Wash spears, cut off woody bottoms. Blanch two minutes, freeze on tray. Separate and store in freezer-safe containers.	Sautéed, cut into pieces and cooked into soups or dishes with sauce (to hide mushy texture when cooked)
Beets	Remove all but ½ inch of stem, parboil until fully cooked. Peel, trim, cut into chunks or slices, freeze in tubs or bags.	As is, warmed and seasoned, mixed with other cooked root vegetables, made into borscht
Berries	Rinse gently, dry thoroughly. Freeze on tray with berries not touching each other and then store in freezer bags.	As is or in smoothies, pies, jams, dessert sauces, toppings
Broccoli	Wash thoroughly. Cut florets off main stems, cut stems into 1-inch chunks and florets into 2-inch pieces. Blanch two to three minutes. Store in tubs or bags.	Chopped in stir-fries; creamed for broccoli soup; hidden in casseroles, stews, or other soups (previously frozen broccoli can look unappealing)
Carrots	Wash, peel, cut into chunks or slices. Blanch two minutes. Freeze in tubs or bags.	Heated and served with butter and ginger or garlic; chopped in stir-fries, soups, or stews; pureed for carrot soup
Cauliflower	Trim stems, cut heads into 2-inch pieces, blanch three minutes. Freeze on tray, store in bags.	Heated and served with butter; in soups and stews; as a thickener for creamy soups
Cherries	Wash, pit, dry. Freeze on trays (not touching each other), store in freezer bags.	In pies, ice cream, sorbet, fruit compotes, smoothies
Corn	Husk, remove silk, wash. Blanch four minutes, dry on cloth. Kernels: remove with knife or tool, store in tubs or bags. Whole ears: cut off tips, freeze on trays, store in freezer bags.	Kernels: use plain or in casseroles, soups, stews, salsa, chowders, or salads; whole ears: drop in boiling water and eat or remove kernels for use

Fruit or Vegetable	Preparation before Freezing	Use after Defrosting
Green beans	Wash, snap off stem ends. Leave whole or cut into 1-inch lengths. Blanch two to three minutes, freeze on trays, store in bags or containers.	Sautéed alone or with other vegetables, steamed, in soups or stews, fried with garlic or onions and slivered almonds
Greens (e.g., kale, chard, spinach, beet/turnip greens, Asian greens)	Wash, remove stems. Blanch two minutes, chop into strips. Store in freezer bags in individual batches.	In soups and stews, in meat loaf, steamed with broth and seasonings
Melon	Peel, cut into chunks. Freeze on trays (pieces not touching), store in freezer bags.	In smoothies, ice cream, sorbet
Onions	Peel, dice or slice. Store in freezer bags.	Sautéed or in soups and stews as if fresh
Peaches	Peel and freeze in slices or halves on trays. Store in freezer bags. Will brown.	In pies, smoothies, ice cream, sorbet, dessert toppings, cobblers
Peas	Shell, blanch two minutes. For looser packing, freeze on tray and then break apart, store in freezer bags. For big chunks of peas, freeze in tubs or bags straight after blanching.	As is, steamed/cooked with onions and butter, in casseroles, soups, stews, meat loaf
Peas: snow/snap	Rinse, remove strings and stems, blanch one minute. Leave whole or cut into 1-inch pieces. Freeze on trays (not touching), store in freezer bags.	Steamed or in stir-fries, soups, stews, salads
Plums	Wash, pit, cut in half. Freeze on trays, store in freezer bags.	As is, with light sugar syrup, or in cobblers, fruit compotes, jams, ice cream, sorbet
Pumpkin and winter squashes	Cut in halves or quarters, bake until soft. Scoop out and drain flesh, freeze in tubs or bags (flatten bags for easy storage).	As is or in pumpkin pie, pumpkin soup, pumpkin bread
Rhubarb	Remove and discard leaves, wash stems, slice into 1-inch pieces. Freeze on trays, store in freezer bags.	In pies (alone or with strawberries), meat sauces, or jams; over desserts as a sauce or in cooked pieces
Summer squash	Wash, slice or cube into $\frac{1}{2}$-inch pieces. Blanch two to three minutes, drain. Freeze in tubs.	In ratatouille, soups, stews (loses flavor and gets mushy when frozen)
Tomatoes	Rinse, freeze on trays, store in freezer bags.	Cook into sauce, use in stews or ratatouille, chop and drain for salsa

Root Cellaring

What to do with that beautiful homegrown bounty when you have too much to eat at once? Because some produce doesn't have to be specially prepared to be stored, you can try a form of root cellaring to save time and your precious harvest.

Why would you want to do this?
You grew produce and have an abundance of it, so it only makes sense to try to store it, and there are some types of produce that you'd rather not pickle or can.

Why wouldn't you want to do this?
You don't have any outdoor space or aren't interested in storing food at all.

Is there an easier way?
This project is about as easy as root cellaring can get; if you have a basement or yard, there are many ways to make a root cellar that is larger and more elaborate (as well as more useful but also more complicated).

Cost comparison:
This saves you the cost of buying root vegetables from the store.

Learn more about it:
Putting Food By (Plume, 4th ed., 1991) by Janet Greene, Ruth Hertzberg, and Beatrice Vaughan is a great all-around book on food preservation, and it has a wonderful chapter on root cellaring, with a variety of different projects, styles, and sizes. *Root Cellaring* (Storey, 2nd ed., 1991) by Mike Bubel and Nancy Bubel is a comprehensive book, and these authors are my DIY role models.

Beets (ABOVE) and radishes (BELOW) are just two of the vegetables that can be preserved by root cellaring.

Root cellars seem like throwbacks to the days of our grandmothers' or great-grandmothers' houses and their big concrete basements with separate rooms for storing canned goods and barrels of apples and potatoes. With grocery-store produce always available, storing fresh produce may seem unnecessary and antiquated, but when you take the time and effort to grow something, it is a waste to not preserve the excess of the harvest. Many easy-to-grow foods can be stored with minimal processing (in the nonindustrial sense of the word) without taking up room in the refrigerator, freezer, or cabinets.

I grew up with a "potato-and-onion cabinet" with a screened bottom that sat over a hole to the basement. The cool air from below came up into the mesh shelved

cabinet, keeping those items cooler than the rest of the kitchen. Airflow is important—without ventilation, mold and rot can build up from increased humidity produced by the vegetables.

When it gets cold outside, people tend to bring everything inside or under cover. Reverse that process and put a small version of a root cellar outside to take advantage of the natural chill for saving your hard-earned garden crops.

Allow ample space around each vegetable.

Materials:

☐ Large plastic tub with a lid or cover (if your winter temperatures routinely stay below 40 degrees Fahrenheit) or large Styrofoam cooler with a lid or cover (if your winters are mild or unpredictable)
☐ Clean construction-grade sand
☐ Cold outside area, such as a deck, patio, porch, or uninsulated garage
☐ Root vegetables, such as carrots, beets, turnips, and parsnips

Step 1: Wash out your plastic tub or cooler well and dry the interior.

Step 2: Moisten the sand. If it is in a plastic bag, poke a fork into the bottom several times and then pour water into the top until it runs out of the holes in the bottom. Let the bag stand and drain for a while—you want the sand to be damp, not soggy. If it is in a paper bag, you can just leave the bag out in the rain for a few days; when you're ready to pack the root cellar, don't move the bag (it will tear and spill) but just cut a slit in it and scoop the damp sand out.

Step 3: Have your vegetables ready and at hand. To prepare them, harvest them in late fall and trim the green tops to a ½-inch length. Wash off the dirt—they don't have to be immaculate, and they can be damp for storage.

Step 4: Put a dense 1-inch layer of moist sand on the bottom of your container.

Step 5: Lay the vegetables on the layer of sand, making sure that the vegetables don't touch each other or

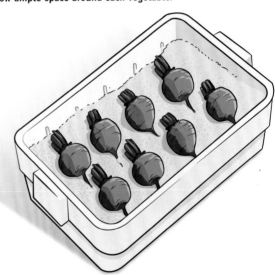

Leave enough space to cover the top layer of vegetables with sand.

the container sides. Fit in as many as you can, keeping at least ¼–½ inch of space around each.

Step 6: Pour sand all over the layer of vegetables, gently packing it in and flattening it until the vegetables are completely covered by about another inch of sand.

Step 7: Repeat Steps 5 and 6 as many times as possible, ending with a layer of sand.

Step 8: Cover the box and store it in a cold, dark spot on your deck or porch. When you use the vegetables, dig out the desired quantity and make sure that the remaining ones are covered entirely. These vegetables should stay fresh for four to six months.

SECTION II:
For You and Your Home

Cleaning Supplies

Do you know what's in the products that you use to clean your house? Commercially made cleaning products can kill germs and wash away dirt, but over the years, many chemicals in them have been shown to be toxic to our bodies. Why not return to the basics, save money, and revel in the freshness of a clean home without harsh chemicals and artificial perfumes?

Why would you want to do this?
You'll know what is in your cleaning products, and you'll reduce your waste by reusing containers.

Why wouldn't you want to do this?
You like the scents of commercially made cleansers, or you don't believe that a homemade product will clean as well as a store-bought one.

How does this differ from store-bought versions?
Your homemade cleansers are natural products. If you prefer scented cleaning products but want to keep them natural, you can use essential oils in your homemade versions.

Cost comparison:
Homemade cleaning products can be made for one-tenth to one-third of the cost of store-bought products.

Skills needed:
Only measuring and mixing.

Learn more about it:
Natural Cleaning for Your Home (Lark Books, 1998) by Casey Kellar (the recipes presented here are among the chemical-free products in Casey's book); *Clean & Simple* (Time Life Education, 1999) by Christine Halvorson and Kenneth M. Sheldon.

You don't need heavy-duty chemicals to get crystal-clear windows.

When my husband and I decided that I would become a stay-at-home parent, I became very conscious about the ways I could save money. For me, that meant learning how to do and make as many things as I could myself. As a new parent, I also became very conscious about our home environment and what I was putting on the floors, counters, tubs, and tiles to clean them. The contact that our toddlers—and their hands and mouths—had with these surfaces was extensive. In the kitchen, even putting our food in the sink and on the countertops ran the risk of chemical contact. (Grocery baggers put cleaning-solution bottles in separate bags from the edibles!)

Cleaning products can be made from basic ingredients that clean effectively without leaving toxic residues in the name of hygiene. Homemade cleaners do not last as long because they are made of perishable ingredients and do not contain artificial preservatives, so you simply make small batches and use them up within a couple of months.

Baking soda, borax, cornstarch, rubbing alcohol, salt, and vinegar are some of the basic ingredients that work great as cleansers, either on their own or in combination with others. You probably already have most of the ingredients that you'll need. The following recipes are excerpted from *Natural Cleaning for Your Home* by Casey Kellar. Copyright © 1998 by Casey Kellar. Used with permission from Sterling Publishing Co., Inc.

Materials for Each Project:
☐ Bottles and containers: rinse and reuse empty bottles from commercial cleaning products—they are usually made of heavy-duty plastic (to contain the toxins that might destroy less sturdy material) and have squirt lids or spray tops
☐ Measuring containers

No-Streak Window Cleaner

Ingredients:
☐ 1 cup white vinegar
☐ 2 Tbsp isopropyl alcohol (rubbing alcohol)

Mix the ingredients together in a spray bottle. Shake well before using. Spray on a window or mirror, and wipe off with a soft, dry cloth or wadded-up newspaper. Store the bottle in a cool, dark place.

Borax is an effective natural mold and mildew fighter.

Liquid Tub and Tile Cleaner

Ingredients:
- ☐ ½ cup ammonia
- ☐ ½ cup white vinegar
- ☐ ¼ cup baking soda
- ☐ 1 tsp borax

Mix the ingredients together and put in a squeeze-top container. Stir or shake well before using. Use with a scrub brush or toilet brush, and rinse surfaces with a damp sponge to remove any cleanser residue.

Heavy-Grime Marble and Granite Cleaner

Ingredients:
- ☐ ¾ cup borax
- ☐ ¼ cup water

Mix ingredients into a paste. Use it with a sponge to scrub surfaces. Discard any remaining paste.

Stainless Steel Polish

Ingredients:
- ☐ ½ cup ammonia
- ☐ ½ cup water

Mix ingredients in a spray bottle. Spray on and rinse off.

Additional Advice
- Label each bottle with its contents and the date on which you made it.
- Play with essential oils to add scent to your cleaning products. Aromatherapy is an ancient practice, and fragrances can help enhance the happy and calm feelings of having a clean house.
- Castile soap can be a good additive. This is a pure form of soap that is available in bar and liquid forms. Some

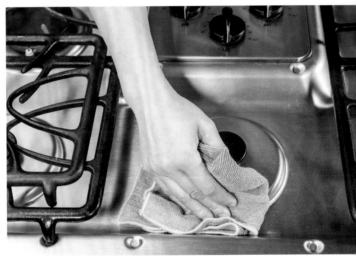

From granite countertops (TOP) to stainless appliances (BOTTOM), your kitchen can be cleaned with natural products.

people (myself included) don't always feel that something is truly clean unless it gets "soapy." While I used to chalk this up to a silly aesthetic, soap in fact cleans well because the soap molecules attach and hold on to oil molecules in a way that water alone cannot. Some things just need soap to get truly clean, and there are many recipes for homemade cleaners that use castile soap.
- Sometimes we assume that a stain or mess needs "big guns" (i.e., harsh chemicals) to be cleaned up properly. However, most messes in the home happen as the result of daily living and don't pose a threat to our health. Natural cleaners get rid of the worst of the germs while preserving the "good" bacterial environment on the surfaces of your home.

Skin-Care Products

When was the last time you read the labels of the products you put on your skin? Pick up a couple of items now. How many of the ingredients can you pronounce, let alone recognize? Do we really need to be putting stuff like that on our skin? If we won't put something *in* our bodies that we don't recognize, why would we put it *on* our bodies?

Why would you want to do this?

It's fun to try, homemade products are better for your body than those loaded with laboratory-derived ingredients, and they make great gifts.

Why wouldn't you want to do this?

You don't want to purchase the few custom ingredients, you are not concerned about what's in your skin-care products, or you don't use skin-care products.

How does this differ from store-bought versions?

Much of what you pay for with cosmetics is labeling and packaging; this is not a factor with your homemade versions. They are satisfying to use, partly because of their wholesome ingredients and partly because you made them yourself.

Cost comparison:

Homemade versions end up costing about a quarter to half as much as the store-bought versions.

Skills needed:

Simple cooking/mixing skills, but use caution melting wax on the stove—both the boiling water and the wax can spatter.

Learn more about it:

The Herbal Body Book (Storey, 1994) by Stephanie Tourles; *Earth Mother Herbal* (Fair Winds Press, 2002) by Shatoiya de la Tour and Richard de la Tour.

Our skin is our largest organ, and it is our first line of defense against harm—from impact, ultraviolet rays, and germs. While many of us take our skin for granted, even those who slather it with expensive cosmetics aren't really treating it fairly.

Many skin-care products are loaded with preservatives. Some contain petroleum derivatives, which clog the pores and make the skin's natural functions of excreting wastes and modulating body temperature more difficult. Some moisturizers contain isopropyl alcohol, which, instead of hydrating the skin, actually serves to dry it out more.

I prefer to put products on my skin that are made from ingredients that I can understand (and pronounce) and that I know come from natural sources that are safe to apply or ingest. Some cosmetic lines carry such products, and many of them are wonderful—but they can be very expensive. Although some of the ingredients I use to make my own cosmetics aren't cheap, per se, just a few drops are all that is necessary for an entire batch, and I can be certain of what is going into my skin-care products.

The US Food and Drug Administration (FDA) regulates many of the additives in medicines and cosmetics, which must undergo extensive testing before they can be deemed safe for human use. This is a good system, but I prefer to use products that have been used for thousands of years in the practices of herbalism. Many medical advances have come from plant compounds. Herbal effects can be slower, but they are also gentler on the body.

Homemade cosmetics can require specialty ingredients, most of which are easily found at natural-food stores and pharmacies or available online. Get small quantities, because quite often a little goes a long way.

I keep separate cooking implements for these recipes, simply because some of the compounds (especially wax or emulsifying agents) are difficult to thoroughly clean away. You can get packs of cheap disposable tin pans (I wash and reuse them) in grocery stores for heating and mixing, and I store my beeswax in a used soup can and melt it in a pan of boiling water.

The recipes here are excerpted from *The Herbal Body Book*, copyright © 1994 Stephanie L. Tourles, used by permission of Storey Publishing LLC, all rights reserved. I hope they will encourage you to try more on your own.

Lip Balm

Homemade lip balms make great gifts—you can either put the finished product in little tubs or refill used-up lip-balm tubes. The essential oil lets you pick a custom flavor, and the honey makes it lip-smackingly tasty. I use a tiny cocktail whisk to mix it up, and then I put it in the containers before it is completely hardened. This also doubles as a cuticle cream.

Materials/Ingredients:

- ☐ Double boiler
- ☐ Small whisk
- ☐ Small storage tubs or jars
- ☐ 2 tsp beeswax
- ☐ 7–8 tsp sweet almond, castor, jojoba, or quality vegetable oil
- ☐ 1 tsp honey
- ☐ 5 drops essential oil of lemon, lime, orange, tangerine, peppermint, or apple blossom

Melt the oil and beeswax together (using 8 teaspoons of oil will result in a thinner, glossier consistency) in a small saucepan over low heat or in a double boiler just until the wax is melted. Remove from heat. Add the honey and blend the mixture thoroughly. Stir the mixture occasionally as it cools to prevent separation. When the mixture is almost cooled, add your essential oil of choice and stir thoroughly. The lip balm should have the consistency of vegetable shortening, such as Crisco, when ready. Store in a small container.

Baby's Bottom Cream

This is a nice, thick, protective cream that is good for preventing diaper rash and for protecting skin in the cold winter air or while skiing.

Materials/Ingredients:

- ☐ Double boiler
- ☐ Small whisk
- ☐ Small storage tubs or jars
- ☐ 2 tsp nonpetroleum jelly
- ☐ 2 tsp cocoa butter
- ☐ 2 Tbsp grape-seed, jojoba, or castor oil
- ☐ 2 drops essential oil of orange blossom, apple blossom, carrot seed, or lemon balm (*Melissa officinalis*)

Heat all of the ingredients (except the essential oil) in a small pan just until the cocoa butter is melted. Remove the mixture from the heat, allow it to cool a bit, and then stir it occasionally until it is cool and thick. Add the essential oil and then stir the mixture again. The cream will be relatively thick and clear. Store it in a shallow tub or jar; it does not need refrigeration. Use 1 teaspoon per application.

Herbal Soap Balls

Castile soap is a very pure, simple, olive-oil-based soap product, and it makes concocting your own cosmetics and cleaning products that much easier (you do not want to play with making soap from scratch, believe me!). It is available at natural-food stores in either liquid or bar form; for this recipe, grate the bars with a coarse cheese grater. These soap balls are fun to make and give as gifts.

Materials/Ingredients:

- ☐ Double boiler
- ☐ Small whisk
- ☐ Coarse cheese grater
- ☐ Wax paper
- ☐ Two 3½-ounce bars unscented castile soap or pure glycerin soap, grated
- ☐ 2 Tbsp ground oatmeal or cornmeal
- ☐ 1 Tbsp crushed lavender, rosemary, or peppermint (or your favorite herb for your skin type)
- ☐ 10 drops essential oil, your choice
- ☐ 1 Tbsp anhydrous lanolin
- ☐ 1 Tbsp sweet almond, castor, jojoba, or quality vegetable oil, plus additional oil for hands to prevent sticking when rolling

Melt the grated soap, lanolin, and essential oil over low heat in a small saucepan or double boiler until the mixture is a very soft, mushy consistency. Stir occasionally while melting. Remove the mixture from the heat and stir in the remaining ingredients.

While the mixture is still hot, oil your hands and form the mixture into balls. You can make the soap balls any size you like, but I think the size of a lime is a good size to handle. Place the balls on wax paper to cool. Use this soap as you would regular soap, but do not leave it in a puddle in the shower, as it will melt. If you prefer round cakes of soap to balls, this recipe will yield approximately two cakes the size of rice cakes.

Insect Repellent

This insect repellent is a natural alternative to chemical sprays. It works best on days when the mosquitoes are only slightly to moderately hungry. If they're voracious, seek a stronger concoction.

Materials/Ingredients:

- ☐ One 16-ounce or two 8-ounce spray bottles
- ☐ 2 cups witch hazel
- ☐ 1½ tsp essential oil of citronella or lemongrass
- ☐ 1 Tbsp cider vinegar

Combine all ingredients in a 16-ounce spray bottle or two 8-ounce spray bottles and shake vigorously. This formula requires no refrigeration and keeps indefinitely. Apply liberally as needed, keeping away from the eyes, nose, and mouth.

Lemon balm oil contains vitamin C and has a pleasing scent.

Castile soap is a gentle yet effective natural cleanser.

Make Your Own Candles

Homemade candles are easy and inexpensive to make, and they are nice gifts to give. I used to hoard candles, saving them for special occasions and not wanting to spend money to replace them. Now we light candles for every dinner and have found, over the years, that it makes our family time quieter, calmer, and cozier because we are embraced in a circle of light and warmth (and the dim light also serves to disguise ingredients in meals that picky eaters would otherwise remove!).

Why would you want to do this?
It is fun, easy, and cheap, and it offers a sense of self-sufficiency.

Why wouldn't you want to do this?
You don't use candles or aren't comfortable working with hot wax.

Is there an easier way?
You can splurge on candle molds, and while they may initially be more expensive than the number of candles you'd buy, you can use them limitless times, making the cost eventually negligible. They come in myriad sizes and shapes.

Cost comparison:
The price of homemade candles is a small fraction of that of their store-bought counterparts.

Skills needed:
None in particular except a steady hand for pouring melted wax into the molds while keeping the wicks upright.

Further refinements:
You can branch out beyond beeswax to other kinds of wax, you can try coloring your candles and scenting them with essential oils, and you can try different styles of candle making (such as hand dipping) to expand your repertoire of shapes and styles. Using beeswax sheets is great for young kids because they can roll or shape the wax around a short piece of wick in almost any style, and voilà— handmade gifts for the relatives!

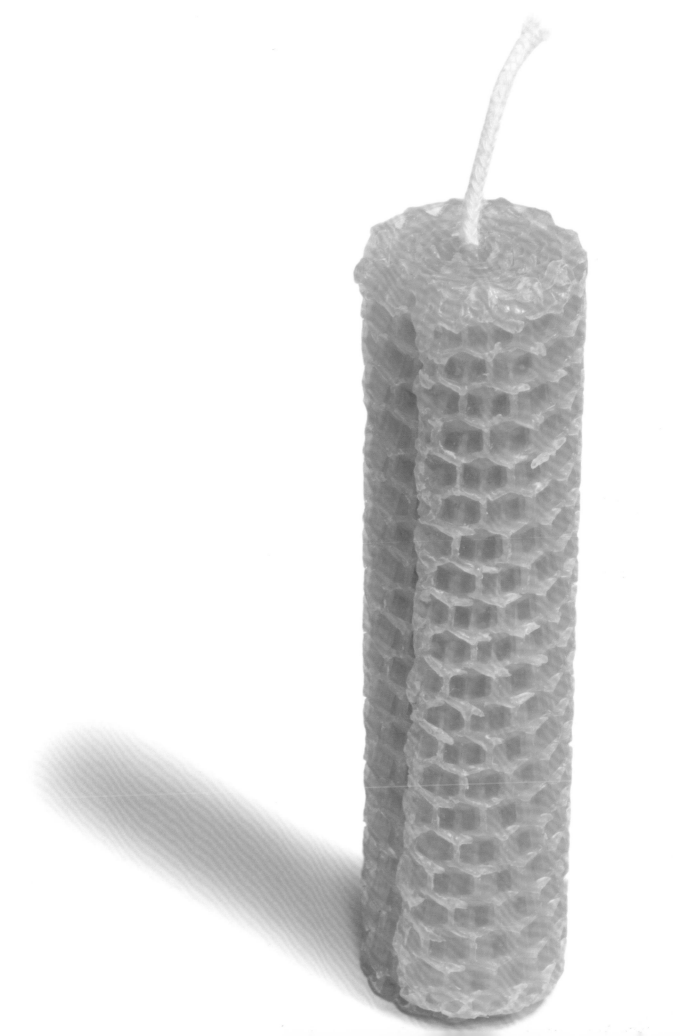

In the following projects, we use two beeswax methods: rolled, with beeswax sheets; and poured, using hot melted beeswax. If you choose to invest in molds for poured candles, you'll see that many have small holes to assist with threading the wick through or inserts that ease the removal of the finished candle.

Rolled Taper

• •

Materials:

☐ **Beeswax sheets**
☐ **Roll of wick**
☐ **Scissors**

• •

Step 1: Take a sheet of beeswax out and let it warm up a bit to make it more malleable. You can let it sit in a sunny area for a few minutes to speed up the process, but watch to make sure that it doesn't melt.

Step 2: Decide on the height and diameter of your candle; cut the sheet according to your desired size, if needed. The sheets of beeswax I use are 8½ inches by 16½ inches, so I make two 8½-inch tapers approximately 1 inch in diameter, depending how tightly I roll them.

Step 3: Lay the wick along one edge of the sheet, pressing it carefully into the wax to hold it in place before cutting the wick. I fold a little bit of wax over mine to secure it. When the wick is secured in place, cut it from the roll, leaving at least half an inch of wick at the top.

Step 4: Start rolling the sheet tightly, working along the length of the taper. It's OK to press down firmly as you roll, as this will also compress the wax and help it stick together.

Secure the wick along one edge of the beeswax sheet.

This illustration is intended to show the basic procedure of rolling the wax around the wick. You will roll yours much more tightly.

Step 5: When the wax is all rolled up, roll the candle a few times back and forth on your work surface to firmly secure the end of the sheet against the finished candle body. You can angle the bottom of the candle inward a bit by smoothing it with your fingers; this will give the candle a tighter, smoother end to tuck into a holder.

Poured Square Column

• •

Materials:

☐ **Quart or pint paper milk cartons**
☐ **Large (28-ounce or larger) metal can with label and one end removed**
☐ **Old saucepan with diameter slightly larger than that of your metal can**
☐ **Oven mitt**
☐ **Bulk beeswax**
☐ **Roll of wick**

• •

Step 1: Rinse out the milk carton and cut off the top so that you are left with a rectangular box.

Step 2: Poke a small hole in the center of the bottom of the carton—just big enough to feed the wick through while keeping a snug fit. Pull through a length of wick that is longer than the height of the carton, leaving a tail sticking out of the bottom; you can tie a small knot in it if you'd like. Set the milk-carton mold on a plate or thick pile of newspapers with the tail of the wick lying flat against the underside of the carton. Try to have the carton as level as possible.

Step 3: Melt enough bulk wax to make a candle of the desired height (see the How Much Wax? sidebar). To do so, put the wax in the empty tin can, put the can in a saucepan filled with water, and boil the water. Be careful—you don't want to get any water in the wax, and the wax can spatter or burn as it melts against the hot sides of the can.

Step 4: When the wax is completely melted, reach into the milk carton and hold the wick upright so it will be straight inside the candle. With the other hand in an oven mitt (or using a helper), carefully pour the melted beeswax into the milk carton while holding the wick

Make a small knot at the bottom (RIGHT), with an inch or less of wick at the top (ABOVE).

gently taut and keeping your hand above the hot wax flow. Be careful not to splash wax as you fill the carton to the desired height.

Step 5: Let the candle cool completely, ideally overnight. The larger and thicker the candle, the longer the time needed for the candle to thoroughly solidify.

Step 6: When the candle is completely solid, tear the milk carton away from the sides and the bottom of the candle. Trim the bottom wick flush with the candle bottom, and trim the upper wick to no more than an inch at the top. You can smooth imperfections or edges with your fingers or a quick dip of the whole candle into a bowl of hot water.

How Much Wax?

CandleTech (www.candletech.com) includes, among its many fine resources for candle makers, an online calculator to help you figure out how much wax you'll need for a given project. You just enter the dimensions of your mold, and CandleTech tells you how much wax you will need. You will find the site's calculators under "General Info" on the home page.

For further reference, 1 pound of beeswax yields approximately 16 ounces of liquid wax, 1 pound of paraffin wax yields approximately 20 ounces of liquid wax, and 1 pound of soy wax yields approximately 18 ounces of liquid wax.

Poured Round Column

Materials:

- ☐ Toilet-paper or paper-towel tubes
- ☐ Aluminum foil
- ☐ Rubber bands or tape
- ☐ Large (28-ounce or larger) metal can with label and one end removed
- ☐ Old saucepan with diameter slightly larger than that of your metal can
- ☐ Oven mitt
- ☐ Bulk beeswax
- ☐ Roll of wick
- ☐ Heat-safe container

Step 1: Prepare your paper-tube mold. Fold a square of aluminum foil so that it is larger than the bottom of the tube and then fold it over the tube's opening. Secure the foil around the sides with a tight rubber band or a strip of tape wrapped around several times.

Step 2: Poke a small hole (carefully—the foil can tear easily) in the center of the foil and feed the wick through the hole and up the full height of the tube. It should protrude out of the top of the tube by at least an inch, with a small tail protruding from the bottom foil.

Step 3: Stand the tube up in a heat-safe empty container that will keep the tube upright and stable while you're pouring the wax.

Step 4: Follow Steps 3–5 of the Poured Square Column, unmolding by tearing off the paper tube. The tube may leave some paper on the candle, which you can peel off or remove by quickly dipping the candle in hot water.

Foil secured with a rubber band forms the bottom of the "mold."

Dryer Bags and Sachets

Why buy, use, and throw away chemical-laden dryer sheets when you can make your own dryer bag and reuse it for many loads of laundry? You can also keep your clothes smelling nice by putting cloth bags full of herbs or cedar shavings in your drawers; the latter will also help fend off moths that eat sweater fibers.

Why would you want to do this?
It is easy, it's cheaper than buying dryer sheets, it can help protect your sweaters from munching moths, and it makes your clothes and your drawers smell good without artificial fragrances.

Why wouldn't you want to do this?
You don't like floral or herbal smells, or you are sensitive to scents.

Is there an easier way?
If you don't want to do any sewing, you can take a knee-high nylon stocking or lightweight sock, fill it a quarter of the way to halfway full with the dried herb of your choice, and either tie a knot in the top or tie a ribbon or string around the opening. It doesn't get much easier than that.

Cost comparison:
Your homemade sachets save money because they last for years and are easy to freshen up when the scent weakens.

Skills needed:
Basic sewing skills and the ability to tie a knot.

Lavender is a popular herb in skin-care products and aroma-therapy because of its soothing properties.

Since ancient times, dried herbs have been used to scent our surroundings and chase away the germs and bugs that often accompany (or are betrayed by) foul smells. In art and literature from throughout history, we find depictions and mentions of scented handkerchiefs, smelling salts, snuff, strewn herbs, wreaths, nosegays, bouquets, and many other ways that herbs were used to spare our ancestors' noses from suffering the scents of daily life. The predecessors of our modern perfumes and colognes worked hard to disguise the smells of humanity because showers and baths were few and far between.

Thanks to modern sanitation, scenting ourselves and our living areas has become more of an option than a necessity—an option we happily choose to exercise. We use herbs and other natural fragrances everywhere in our daily lives to create pleasing aromas.

Even with our laundry, we are not satisfied with just removing bad odors. We want the newly washed clothes to come out of the dryer with an even fresher smell. However, if there is any fragrance on what comes out of my dryer, I like it to be natural.

I have a little cloth bag that I filled with lavender blossoms about three years ago. I haven't refreshed it once, but every time I use it in the dryer with our sheets, it gives off a strong, sweet lavender smell. Roses, lavender, and other floral herbs can leave their scent lingering in fabrics just by proximity, and cedar will help repel the little moths that eat holes in wool, even though you may never see these small offenders. (I never have, but the holes kept appearing until I started using the cedar!)

The lavender in the following instructions can be replaced with whatever dried herb you like, such as rosemary or mint. You can buy bags of cedar shavings at a pet-supply store, or you can make them yourself by getting an untreated piece of cedar 2x4 and either planing it or grating it lengthwise with a heavy-duty grater (such as a secondhand large-holed cheese grater) or rasp.

Lavender Dryer Bag

Materials:
- ☐ Light cotton fabric, size depending on desired finished size
- ☐ Sewing machine or needle and thread
- ☐ Loose dried lavender, approximately 1 cup
- ☐ Optional: Ribbon or string

Step 1: Cut a rectangle of fabric to be twice as long as the finished bag you'd like (e.g., for a 3-inch by 4-inch bag, cut a rectangle of 3 inches by 8 inches).

Step 2: Fold the rectangle in half, right (patterned) sides together, and sew the two longer sides closed. Take the last open side of the bag and fold the fabric over to create a small hem (this is the mouth of the bag).

Step 3: Turn the bag right side out.

Sew the opening closed after you've filled the bag with lavender.

Step 4: Fill the bag approximately three-quarters full with lavender, loosely placed, not packed in.

Step 5: Sew the open mouth closed either on a machine or with a hand whip stitch, or tightly wrap a piece of string or ribbon multiple times around the mouth and then tie it to hold it closed. A secure closure is especially important because the bag will go through the dryer, and you don't want the lavender to spill out.

Keep your bag near the dryer and throw it in any time you want a gentle lavender scent imparted to a load. I made my bag several years ago and use it approximately once a week, and it still leaves a gentle aroma. The more often you refill the bag (if you've used a tie closure), the stronger the scent will be. You can also "refuel" the dried blossoms by applying a few drops of lavender essential oil to the bag.

Lavender Sachet Bag

Materials:
- ☐ **Light cotton fabric, size depending on desired finished size**
- ☐ **Sewing machine or needle and thread**
- ☐ **Ribbon or string**
- ☐ **Loose dried lavender, approximately 1 cup**

The sachet bag is tied closed with a ribbon or string.

Steps 1–4: See the Lavender Dryer Bag project.

Step 5: Tie a pretty ribbon around the mouth end of the bag, closing it tightly. Tie the ribbon into a nice bow and add decorations if desired.

Put your sachet in a clothes drawer or hang it from a hanger in your closet to impart the scent to the surrounding clothes and area.

Cedar Sachet Bag

Materials:
- ☐ **Light cotton fabric, size depending on desired finished size**
- ☐ **Sewing machine or needle and thread**
- ☐ **Cedar shavings, approximately 1 cup**
- ☐ **Optional: Ribbon or string**

Follow Steps 1 through 4 of the Lavender Dryer Bag instructions but make the bag larger (mine is 6 inches square) and substitute cedar shavings for the herbs.

Keep the cedar sachet in areas where you store your woolen sweaters or suits. Moths seem to find the woolen goods and eat holes in them no matter how contained they are.

Sachets keep pests away and impart a pleasing scent.

SECTION III:
Backyard Projects

Build Your Own Rain Barrels

Although we live in a very wet climate, we still have several months each year during which water is considered a scarce commodity. This dry season coincides with the time when plants need the most water for their optimal growth. Using rain barrels to collect water against times of need not only saves money but also provides you with water that is better for your plants. Rainwater is free of the additives and chemicals used in municipal filtration processes, and it takes only one good rainstorm to fill the barrel back up.

Why would you want to do this?
To save money on your water bill and to provide your plants with a chemical-free resource.

Why wouldn't you want to do this?
You don't have room for a rain barrel or don't have plants that need supplemental water.

Is there an easier way?
You can buy a premade rain barrel, but that can be much more expensive than constructing one yourself. The second easiest way is to use a kit to turn your own barrel into a rain barrel, but again, you'll spend more than you probably need to.

Cost comparison:
The materials for a homemade version cost less than half of what even an inexpensive prefab rain barrel costs. Try to find the 55-gallon drum secondhand; warehouses often give away food-grade barrels or sell them for a small fee.

Skills needed:
Basic construction know-how—drilling holes, cutting off a downspout if necessary, installing simple plumbing pieces.

Further refinements/learn more about it:
You can find filtration devices that clean the collected water, and many people have gone so far as to collect rainwater in cisterns for use in their home plumbing systems. The following resources can help you get started: *Water Storage* (Oasis Design, 2005) by Art Ludwig; *Rainwater Collection for the Mechanically Challenged* (Tank Town, 2005) by Suzy Banks and Richard Heinichen; RainHarvest Systems, www.rainharvest.com.

You can cut a hole in the top, if needed, for the downspout.

There are several issues to consider before you begin this project. First is the opaqueness of the barrels you use. I bought white food-grade 55-gallon drums, and although they seem opaque, enough sunlight can get through them to encourage the growth of algae in the water. This is not harmful to the plants that I water, but it can get smelly and plug up the spigot. My solution was to paint the barrels black with special plastic-surface spray paint, but I advise getting colored barrels, as dark as possible, if you can.

The second issue is how you will deal with possible overflow. You can likely run your downspout directly into the top of the barrel because most barrels have big screw-top openings. (If your downspout is too large or the wrong shape to fit the opening, you can cut a larger hole.) Be aware, though, that a heavy rainstorm can fill a barrel in just a couple of hours, and the barrel will overflow if the water has no outlet. Drilling a hole in the side of the barrel at the top will control the overflow, but water spilling out of the barrel may cause problems—problems that the downspout is supposed to solve in the first place.

There are two easy solutions: One is to use a diverter, a device that is inserted into the downspout on one end and into the barrel on the other; it diverts water from the downspout into the barrel until the barrel is full, then reroutes the water back down the downspout. The other is to drill the upper hole in the side of the barrel but instead of leaving it open, run a hose from it to wherever you'd like the extra water to go (ideally, away from your foundation). This upper hole can also be used to link several rain barrels with hose pipe so that when one barrel is full, the water will run into an adjacent barrel.

The third issue to consider is location of the barrel. A rain barrel is a rather large piece of outdoor "furniture" and not terribly attractive (if appearance matters to you). It will obviously need to go next to or under a downspout, and therefore against an outer wall of the house, probably near a corner. Make sure that the ground underneath the barrel is solid and level—once the barrel is full, you will not be able to move it. A gallon of water weighs more than 8 pounds, so the barrel will weigh about 500 pounds when full, and you do not want it to sink, lean, or fall over. You can imagine what would happen!

Finally, think about how you want to access the water in the barrel. If you water your plants by hand, you'll need to be able to fit a watering can under the bottom spigot. In this case, consider raising the barrel up on blocks or some sort of platform for easy access to the spigot. Whatever you place the barrel on should be strong and stable. If you just want to run a pipe or hose straight from the spigot, you might be able to sit the barrel directly on the ground—but only if the water's destination is lower than the barrel's bottom. Keep in mind that water only flows downward—if the distribution end of your hose is higher than the bottom of the barrel, the water won't flow out. I've made that mistake, and, believe me, you do not want to be standing in your yard with a full rain barrel and a hose that's bone dry. Now, let's get started!

Materials:

- ☐ **Plastic food-grade 55-gallon drum (metal drums eventually rust and are more difficult to deal with)**
- ☐ **Threaded hose spigot**
- ☐ **Drill, with bit slightly smaller than the spigot threading**
- ☐ **Hacksaw or fine-toothed saw**
- ☐ **Optional: Rubber washer for spigot**
- ☐ **Optional: Gutter/downspout diverter or diverter kit (nice to have)**
- ☐ **Optional: Bricks or cinder blocks**

Step 1: Drill a hole in the side of the barrel, approximately 3–4 inches up from the bottom. The hole should be slightly smaller than the diameter of the threads on your hose spigot to ensure a tight fit when you screw the spigot in.

Step 2: Place the barrel where it will live.

Step 3: Drill a hole in the side of the barrel approximately 2–3 inches down from the top. Decide exactly where to put this hole based on where the barrel will sit in relation to the downspout and where you want the spigot to aim. If using a diverter, make the hole in accordance with the size of the tube that you'll attach between the downspout and rain barrel. If you'll be using a hose in this hole for overflow protection, the hole should be the right size to insert the hose connection snugly.

Step 4: Mark where to cut the downspout. If you are putting it directly into the barrel, it should extend a few inches below the top of the barrel. If you are using a diverter, make sure the bottom of the downspout is level with or slightly above the height of the upper barrel inflow hole. Use your saw to cut the downspout to the proper length.

Step 5: Place the barrel on its side, and screw the spigot into the lower hole. You can put a rubber washer on the spigot so it will be squeezed between the barrel and tap when the spigot is completely inserted. Try to do this step only once, as threads will be carved into the plastic by screwing the spigot in, and it will never be as tight as during the first insertion. Make sure that the spigot is oriented in the proper direction when you do the final turns.

Step 6: If using a diverter kit, follow its directions to attach the diverter to your downspout.

Step 7: Level and turn the barrel to get it into position. Now is the time to make adjustments, before the barrel fills up. Insert the downspout or attach the diverter tube barrel.

Step 8: Wait for a rainstorm.

Additional Advice

- If you want to link multiple barrels, drill a second hole in your barrel, a few inches down from the top, on the side where the next barrel will sit. Connect the barrels with a short length of pipe or hose between those two holes—when the first barrel is full, the water will overflow into the second. Be sure there is a spigot on the bottom of the second barrel as well. If you want to have only one outflow (spigot on just one of the barrels), link your barrels with pipes at the barrel bottoms, just slightly higher than the spigot height. They'll fill at the same time and drain simultaneously.
- If you find that the standing water in your barrels is encouraging mosquitoes, you can float "mosquito disks" on top of the water; these disks release a biopesticide that kills mosquito larva.
- If you have small children, be certain that there is no way for them to climb onto or fall into your barrels. When full, a rain barrel can be as dangerous as a swimming pool for small children, perhaps even more so because barrels are not placed in focal points in the yard or watched as closely.
- If you live in a region with cold winters, disconnect the diverter and empty your rain barrel before freezing temperatures occur. Another option is to drain your barrel partway to allow for the water to expand when it freezes. If it does not have space to expand, the water can cause a plastic barrel to bulge and bow, which can knock it off balance and take the downspout and gutter with it. The barrel will regain roughly the same shape when the water thaws, but it may need to be rebalanced. Metal barrels expand irreparably, often bursting at the seams.

Attach the spigot into the lower hole, ensuring a tight fit.

Build a Bird Feeder

You may not know it, but songbirds are essential to our lives. They keep many bug populations in check by foraging and eating all kinds of insects, and they act as pollinators in their own way. Many have adapted to human intrusion and manage to live and thrive among our buildings and paved roads. Why not help them do the work that helps us? Putting out seed feeders supplements their diets while thanking them for reducing the number of mosquitoes, house-flies, and other nasties that we dislike.

Why would you want to do this?

Birds are not only useful, they also are beautiful and fun to watch.

Why wouldn't you want to do this?

You don't want to have to buy birdseed or don't like creatures hanging around your windows. Some people don't like that spilled birdseed attracts squirrels and chipmunks (who are actually helping by cleaning it up!); others complain that seed lures rats.

How does this differ from a store-bought version?

This project may behave differently than a store-bought model because there is actually quite a bit of engineering that goes into bird feeders, including details that help the seed fall and flow properly and contain the seed while keeping it accessible (you may wonder how birds ever fed themselves before we came along!).

Cost comparison:

You're mainly using objects that you already have around the house, so a homemade bird feeder is practically free.

Skills needed:

Basic handling of a knife and a saw. This can be a good project for teaching kids how to use these tools properly.

Worm Bins

For those who don't have the ground space, the volume of materials, or the interest or ability to maintain a compost pile, worm bins (also known as vermicomposting bins) are a great alternative. Worms do a lot of the work in a cold compost pile anyway, so having a container in which they live and do their work fits well with the urban-farm lifestyle.

Why would you want to do this?

Simply put, less garbage is good for the earth. And why spend money buying compost when you can use your own kitchen scraps—which you've already paid for—to make compost at home? Plus, if you have kids, a worm box is a great educational tool.

Why wouldn't you want to do this?

You're squeamish about worms, bugs, or occasionally dealing with rotting food.

Is there an easier way?

If you want to spend the money, you can order ready-made worm bins, which are very basic but can look nicer than my plastic-tub version.

Cost comparison:

If you can't get the worms for free, they will likely cost more than the plastic tubs, but your total cost will still be cheaper than that of a prefab worm bin.

Skills needed:

Patience and willingness to put up with possible failure (which can be yucky) until the system balances out.

Learn more about it:

Worms Eat My Garbage (Flower Press, 2nd ed., 1997) by Mary Appelhof will inspire anyone to become a worm composter. It includes many different styles of worm boxes, including a coffee-table model that renders moot any protests about not having room to compost. *The Worm Book* (Ten Speed Press, 1998) by Loren Nancarrow and Janet Hogan Taylor is very instructive and helpful when dealing with imbalances in the box.

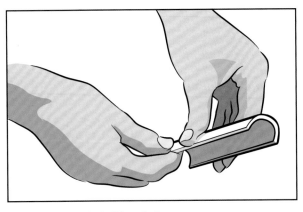

The PVC pipe, cut in half lengthwise.

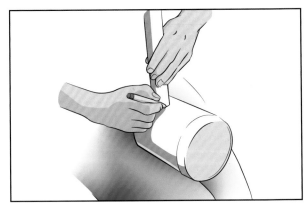

Trace the sideways "D" shape on one side of the container.

Mark the other side, looking through the plastic to line it up.

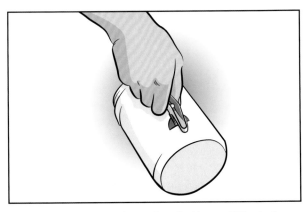

Punch a hole centered underneath each sideways "D" cutout.

Each side should look like this, with the holes lined up.

Stick the PVC pipe through both D-shaped holes.

Mark and cut the pipe to leave a few inches.

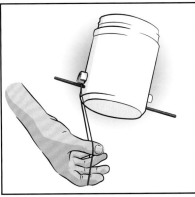

Secure the chopstick perch.

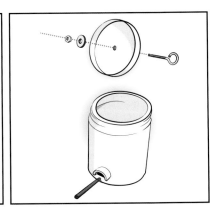

The screw eye creates a hanger in the lid.

Build a Bird Feeder **105**

Bird feeders can be made out of many materials—a bird feeder can be as simple as a large pinecone smeared with peanut butter, rolled in birdseed, and attached to a piece of ribbon or yarn. Both of my children made pinecone birdfeeders as Mother's Day projects in preschool, and the painstakingly handwritten "I Love You, Mommy" on the cards made me cry both times.

Songbirds bring interest and fun to otherwise still, snowy days. Their bright colors stand out against the monochromatic winter landscape, and they are reminders that the warmth of spring always follows the cold.

Store-bought bird feeders are surprisingly expensive, and you can let your imagination run wild and build bird feeders relatively easily out of basic materials. You can salvage the materials for a feeder, quite often from your own kitchen recycling bin. My husband voiced his dismay about the plastic 2-liter-bottle feeder hanging from our Japanese maple, so if you'd prefer not to showcase your recyclables, you can find what you need for many simple feeder styles at a local hardware or thrift store. For this project, I avoided the 2-liter bottle, but any plastic bottle or jar will work.

Materials:

- ☐ Plastic container with a screw-top lid, wide mouth preferred
- ☐ 1-inch- or 1½-inch-diameter PVC pipe longer than the diameter of the container
- ☐ Pencil or marker
- ☐ Saw
- ☐ Utility knife or X-ACTO knife
- ☐ Single-hole punch or drill
- ☐ Chopstick
- ☐ Two lightweight rubber bands
- ☐ Screw eye with washer and nut
- ☐ Plant hook or ceiling hook (if placing under cover, such as a porch) for hanging

Step 1: Clean the container and remove any labels.

Step 2: Cut the piece of PVC in half lengthwise. You will use one of the halves to form the seed holder.

Step 3: Lay the jar on its side on a counter, and then use something to hold it in place (e.g., a heavy book or brick on either side). Take a piece of the cut PVC pipe and place the end against the side of the jar, about a quarter of the way up from the bottom of the jar. Trace the outline of the pipe onto the container (you'll end up with a sideways capital "D"). Turn the jar over 180 degrees and repeat in the same position.

Step 4: Cut the traced shape with the utility knife or X-ACTO blade. Cut it a little smaller than you traced it. This is important because you want the pipe to fit snugly in this hole.

Step 5: Insert the hole punch into the D-shaped cut and punch a hole in the plastic beneath the cut, about ½ inch down and centered. Repeat on the other side. This is the hole for the chopstick perch you'll insert in Step 8.

Step 6: Slide the PVC halfway through the D-shaped holes. Mark the ends that stick out and then remove the pipe. Cut off the ends a few inches out from your marks (you want the pipe to be longer than the container). Slide the pipe back in.

Step 7: Poke or drill a hole centered in the top of the feeder. Screw the screw eye through the hole and attach the washer and nut to the screw eye on the inside of the feeder.

Step 8: Poke the chopstick through the hole-punch holes, centering it to form a perch on either side of the container. To help keep the perch in place, twist a rubber band onto each end of the chopstick, flush against the container.

Step 9: Fill the feeder with birdseed and hang outside. Depending on how secure the lid is and how tightly the screw eye fits into the hole, you may want to hang the feeder under cover so water doesn't get inside and spoil the seed.

Additional Advice

- You can use an old wok, turned upside down with the handle removed and a hole drilled in the center, as a roof/squirrel baffle over any bird feeder.
- Another possibility for a baffle is a wide bowl—metal is best because squirrel claws can't get a grip on it. I salvaged an old deep metal bowl, drilled a large hole in the bottom of it, and mounted it upside down under the feeder on the pole of a hanging feeder stand.

You'll see soil-like matter (the worm's castings) begin to form around the food materials.

In a natural, outdoor environment, compost happens as the result of millions of organisms working together to break down biodegradable matter. While all of the organisms are helpful and contribute to the process, foremost among the workers is the humble worm. Composting with worms (vermicomposting), in essence, capitalizes on nature's methodology and is quite simple, given proper conditions.

Did you know that there are a variety of different worms, and the one that helps in composting is not the well-known night crawler, or earthworm (*Lumbricus terrestris*), of fishing-bait use? The best compost worm is the red wiggler (*Eisenia fetida*), not to be confused with the red worm (*Lumbricus rubellus*), an earthworm relative.

One of the reasons for calling attention to this difference is where the worms live and what they prefer to eat. Earthworms inhabit—yup—earth, in the lower reaches of native soils, and they prefer to eat things found on top of the soil. They like their burrows undisturbed and will crawl up to the top and pull surface items down below. Red wigglers don't mind having their burrows disturbed, which is important in composting because you'll be mixing the material regularly. They will readily eat a wide variety of decaying vegetable matter and can eat their own weight in food daily. Red wigglers are prolific in reproducing, too, possibly doubling their numbers in sixty to ninety days. See if you can find a friend with a compost pile who'll collect some worms for you; otherwise, you can buy them from a nursery or online (be aware that they're not cheap).

Because worms have a tendency to travel, it is best to keep them contained in bins for composting. The bins must allow air circulation, and because worms like cool, moist conditions, you must not let the bins dry out or overheat.

Worms can be fairly finicky eaters. They don't do too well with larger or coarser materials, such as cornstalks or pits. Mine do not like any onions whatsoever, and they tend to avoid citrus. I've spoken to other composters who say that their worms love oranges but won't go near potatoes. I've read that worms love coffee grounds in modest amounts, but they appear to eschew tea bags.

Other bugs and creepy crawlies can, and usually do, appear in a worm bin—tolerance depends on your level of comfort. All of the creatures that find their way into a worm bin are decomposers on some level and therefore beneficial.

Materials:

- ☐ **Two plastic stacking containers or tubs with snug-fitting lids, 14- or 18-gallon size (they can be any shape as long as they leave a little space between the bottoms when stacked)**
- ☐ **Drill with drill bits**
- ☐ **Newspaper**
- ☐ **Red wiggler worms**

Step 1: Prepare your materials. If you're using new tubs, wash them out to get rid of that plastic smell. Shred the newspaper into thin strips of any length, either by hand or with a paper shredder, but keep a few sheets of newspaper intact for later. Have a supply of kitchen scraps and your worms on hand.

Step 2: Turn over one of the plastic tubs and drill holes all over the bottom of it. Holes from $\frac{1}{8}$ inch to $\frac{3}{16}$ inch are best, and you'll need to make many of them. Be sure to put plenty of holes in any areas where liquid could settle.

Step 3: Stack the drilled container in the solid one.

Step 4: Moisten a whole sheet of newspaper and lay it flat on the bottom of the container, covering the holes.

Step 5: Wet a generous amount of shredded newspaper in a bucket or bowl of water, and then wring it out and scatter it loosely over the flat sheet of newspaper. Fill the bin a quarter to half of the way full, loosely with the shreds; this is the start of the worms' bedding. You can also add dead, dry leaves or some light, fluffy dirt. The worms like cool, dark places to hide.

Step 6: Scatter the worms on top of the shredded paper. Watch them a bit—they're cool creatures, and their movements can be mesmerizing.

Step 7: Scatter the kitchen scraps around where you've put the worms. Don't dump the scraps right on top of the worms; you don't want to smother them.

Step 8: Put another moist newspaper sheet on top of everything and put the lid on the bin.

You're done! Keep the bin in a cool environment where you'll be able to check on it regularly—outside is good, because it will attract fruit flies indoors. In the summer, find a shady outdoor spot, out of direct sunlight. In the winter, find a place where it will not freeze. The compost acts as insulation of sorts in cold weather, but freezing temperatures will kill the worm population.

Step 2: Drill holes in the bottom of the container.

Step 5: Add shredded newspaper to the bin.

Now comes the hard part: you've got to be patient and give the worms exactly what they need to create their little ecosystem. They don't like to be overfed, so wait to give them more scraps until you start to see dark spots that look like coffee grounds forming in the food (about a week or two). These spots are the worms' waste, or *castings*. Worms eat our food and excrete a very rich soil amendment, so the more stuff that looks like granular dark soil, the more comfortable the worms are. They are starting their settlement.

If you are patient and careful with the early stages of setting up your bin, you'll be amazed at how efficient

The worms' hard work pays off for you and your garden.

You'll know that the worms are settling into their surroundings when you see their castings, or waste.

As the worms do their job, you'll see more soil-like material than food in your bin.

and helpful the worms can be once it's up and running, turning your waste into garden gold that your plants will love. Worms are still some of my favorite pets.

Additional Advice

- The worm bin must stay moist, but not wet. The lower tub will collect extra liquid and should be emptied periodically. The liquid is called *compost tea* and makes a great additive to watering cans for potted plants.
- Don't let the level of liquid in the bottom bin reach the bottom of the upper bin. Worms can drown. You can attach a spigot to the lower box if you'd like, but it's not hard to just remove the upper box and empty the lower. Be careful, though, because if the liquid leaks or spills, it can stain.
- Keeping a moist layer of paper on top of the food waste will discourage fruit flies and minimize odors. It will also help regulate the moisture within the box. A light layer of damp newspaper is good, or you can use a wet burlap sack that has been wrung out. Pull it back to dump in the scraps, and then replace it to cover them. Sprinkling water on top of this cover will keep it moist and soften it, but don't worry if it dries out periodically. It will usually mold and eventually rot; you can leave it in place and put another layer of the same material right on top. I've used plastic-weave

landscape cloth, which is great because you don't have to worry about keeping it moist (it holds moisture underneath it), and it doesn't rot. It also can be hosed off. If you use a nonrotting material, make sure that it is permeable. Worms can suffocate.

- Worms occasionally try to escape. Sometimes they'll crawl down and drown (hence the need for the bottom paper layer), but usually they'll crawl up the walls of the tub. This can indicate that something isn't right. Check if the materials are too wet or too compacted. Worms like air circulation around them and their food.
- Once the worm population is thriving, try to keep the material inside mixed up as you notice the dark, soil-like castings appearing. When you remove the bin's lid, fold back the inner cover and gently stir the mixture of food and newspaper bedding, trying not to tear the lowest layer of paper. This will also help you see how much of the food the worms are consuming and give you an idea of when it's time to add more.
- In theory, the worms will migrate to where the food is. As your population grows and you add more food, you can alternate sides of the tub. This helps when it's time to take out the finished compost (when you can't see anything that's recognizable as food on one side of the tub, it's time to remove the compost from that side.)

Compost Piles

Many city dwellers have seen their trash-collection costs go up, and some of this is from cities' charging residents a fee for collecting kitchen and yard waste, which the cities then compost themselves. They sell that compost back to gardeners, who use it to amend their soil. I say skip this middle step—getting rid of your kitchen and yard waste and then purchasing it back in finished form—by doing your own composting. There are many different ways to compost, and plenty of places where you can do it. It won't be stinky or unsightly if maintained properly, and you may even be able to do it without anyone's noticing.

Why would you want to do this?
You may save money on your garbage collection, and you will not have to buy compost; also, smaller-scale composting generates less methane than industrial-scale composting does.

Why wouldn't you want to do this?
You haven't got a yard, or you share an outdoor area with people who object to a compost area.

Is there an easier way?
You can purchase a composter, but they aren't always cheap. Keep in mind that much of your success will depend on your commitment to the process as well as the amount that you compost, so give composting a try before you buy.

How does this differ from the store-bought version?
Compost "manufacturers" often add material (such as ground-up wooden pallets) to make a more consistent, but less potent, product.

Cost comparison:
A pile and the finished compost is free; the cost of the compost bin will depend on the materials.

Skills needed:
Nothing in particular other than your attention.

Learn more about it:
Let It Rot! (Storey, 3rd ed., 1998) by Stu Campbell (a classic); *Compost This Book!* (Random House, 1994) by Tom Christopher and Marty Asher; *Easy Composters You Can Build* (A Storey Country Wisdom Bulletin, 1995) by Nick Noyes; *The Rodale Book of Composting* (Rodale Books, 1992), edited by Deborah L. Martin and Grace Gershuny.

Leaves, kitchen scraps, and the like are referred to as *greens*.

For this project, we'll build a contained compost pile with fencing and posts, but you can also use shipping pallets or lumber (new or salvaged) to build a stylish enclosure. Alternatively, you may be able to purchase a yard compost container from your municipality for a nominal fee, or you can salvage 55-gallon drums. You can even just pile the waste in a heap in the corner of your yard. Try to salvage materials until you really know what you want and where you want to put it. Composting is a fairly individual "art," and you have to experiment to find a system that works for you.

There are, in essence, two styles of compost piles, hot and cold (*vermicomposting*, a type of cold composting, is discussed in Project 3, Worm Bins, in this section), and they both involve natural chemistry. The basic formula for compost is: fuel + air + moisture + time = rot. The better you get at composting, the more you can experiment.

Cold Compost Piles

Cold composting, also known as *regular* or *long-term* composting, involves collecting biodegradable matter in one area and letting it sit for an extended period of time. *Biodegradable* basically means "this rots and will make compost." Rain, cold, sun, worms, bugs, and nature's decomposers eventually break down the organic matter, turning it back into soil. If you have a small corner of your yard that you don't regularly use, it could be ideal for a backyard compost pile.

You can put a variety of waste from your house and yard on the pile: weeds, wilted flower bouquets, fallen leaves, dead houseplants, kitchen trimmings (excluding any animal products), cat litter (minus the feces), and dryer lint. Make sure everything gets really wet once in a

while, and in a year or two or three, you can rake everything out of that corner; remove any remaining branches, sticks, and avocado pits; collect the lovely black dirt; and put it where you want new plants to grow. You can then continue to use this area for a compost pile because the decomposer insects and microbes will already be present, or you can plant in the rich soil that remains and start a new pile elsewhere. Composting doesn't get easier than that.

The downside of cold composting is that it does not kill off many of the seeds and pathogens found in your compost items. So depending on where you use your compost, you may have a lot of weeds (or squash or cucumbers, depending on your diet). Plants grown in the compost might have fungi or other soilborne diseases that were not eradicated during decomposition. These problems are rare—and worth the risk, in my opinion, for the small amount of work a cold pile requires.

Hot Compost Piles

Hot piles are more labor-intensive than cold piles, requiring more attention and more detailed construction, but they reward your labor with usable compost faster, and the finished compost is a "cleaner" product.

Hot piles create compost faster because the bacteria responsible for breaking down the materials find a hospitable living environment. Their lives consist of eating and increasing their population. Biodegradable materials that are high in carbon—such as paper, autumn leaves, straw or dry grass, cardboard, wood shavings, and sawdust— are colloquially called *browns*. Materials that are high in nitrogen are referred to as *greens* and are usually the wetter ingredients—kitchen scraps, rotting vegetable matter, fresh grass clippings, animal manure, and the like. Carbon consumption (oxidation) gives the bacteria energy, while nitrogen consumption yields the protein they use to multiply. All of this eating breaks down the matter in the pile as the oxidation generates heat as a by-product.

When the pile cools down, turning it (aerating) causes the process to repeat, reinvigorating the bacterial population with fresh food—thus turning hot compost regularly will help accelerate the decomposition. A hot compost pile can be turned three or four times before the process is complete.

When you build a compost pile, your ultimate goal is to balance the browns and greens. If there are too many browns in a hot pile, the pile won't heat up because it will be too dry. Too many greens, and the pile becomes smelly and slimy from too much moisture. It requires a certain amount of both types of material—in chemical makeup and quantity—to be successful. While a too dry pile will be a letdown, a too wet pile can be a stomach-turning challenge to rectify.

Materials:

- ☐ **12 feet of 3-foot-tall, 1-inch by 1-inch metal fencing**
- ☐ **Four 3-inch lengths of inexpensive 2x4, preferably rot-resistant**
- ☐ **3 small eye hooks**
- ☐ **Staple gun or pound-in U-style nails and a hammer**

For either hot or cold composting, you can build a container, or bin. A bin provides a structure into which successive additions to a cold compost pile are confined tidily and easily. For hot piles, one or more bins help retain heat longer and make the necessary turning easier. Depending on how much compost you plan to make, and how often you think you'll be using it, a series of bins also doubles as storage for the finished product.

Step 1: Lay out your fencing flat on the ground. Staple or secure one of the 2x4s along one end, preferably the end without the wires sticking out, so that the 2x4 is under the fencing.

Step 2: Measure (or count the squares) over 3 feet, slide another 2x4 under the wire, and staple or nail it straight up and down to the fencing. Repeat this twice more, until all four wooden supports are attached. Only one end of the mesh will be attached to wood; the other end will be loose.

Step 3: Turn the whole thing over so that the 2x4s are now on top of the fencing. Standing on one of the wood pieces 3 feet in from an edge, pull the fencing up toward you to bend it at a right angle around the corner of the wood. Repeat on the other edge and middle pieces. If you prefer a more rounded shape, skip this step.

Step 4: Stand the apparatus upright. Screw the eye hooks into the wood piece on the end of the fencing at even intervals.

Step 5: Bring the loose fencing end around to close the apparatus into a square. Bend the wires closest to

Place your bin in an out-of-the-way area that still gives you easy access for tending to it.

the eye hooks down, creating hooks to attach into the screw eyes. This will secure the enclosure.

Step 6: Put the bin in your composting area.

Basic Composting How-To

For a cold pile, simply place all of your materials inside the compost bin you just built and let the pile sit. For a hot pile, gather the materials that you will put onto the pile. Fairly small pieces are best because they'll break down more evenly and faster, but this will happen regardless, as long as the green-to-brown ratio is met. Getting the right balance will take some trial and error. Robert D. Raabe, professor of plant pathology at the University of California Berkeley ("The Rapid Composting Method" http://vric.ucdavis.edu/pdf/compost_rapid compost.pdf), recommends that the compost material has a 30:1 carbon-to-nitrogen ratio for the most effective composting. He writes that this is hard to measure but advises the following: "...experience has shown that mixing equal volumes of green plant material with equal volumes of naturally dry plant material will give approximately a 30:1 carbon-to-nitrogen ratio."

Stack the compost material in alternating layers of greens and browns, trying to keep the layers about 6 to 8 inches thick; don't worry about precision—you can eyeball it. Water the pile after every 4 to 6 layers if it's not already wet. Try to build the pile to be at least as high as it is wide

(this is where a contained system helps). End with a brown layer to prevent odors, flies, and scavengers.

At some point during the first three to four weeks, turn the pile. To open the bin up for turning, unhook the fencing from the eyelets, opening just that panel. Shovel or fork everything out of the bin. Hook the door panel closed again and then shovel or fork the materials back into the bin. Water the pile again after turning and wait another three or four weeks.

Repeat the turning-and-watering process and wait another three to four weeks. At this point, the pile should look pretty well broken down and more like compost. Now you can either turn the pile again and wait some more or use the compost as is.

The following is another approach to hot composting that I found in an *Organic Gardening* magazine from the early 1970s. The fourteen-day, or University of California, method works on piles of a cubic yard or more. It requires more attention and management, but it produces finished compost faster. Here's how it's done:

Day 1: Gather your materials, shred or grind them if needed, place them in a pile, and water well.

Day 2: Check the temperature of pile, which should be around 110 degrees Fahrenheit (long compost thermometers can be purchased from gardening centers or catalogs). Check the moisture of the pile; water if needed.

A fresh layer of greens.

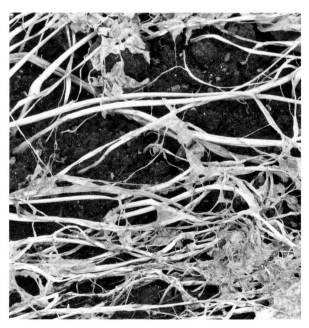

A close look at materials breaking down on the pile.

After two or more weeks, you will see remnants of your materials among the dark, rich compost.

Day 4: Check the temperature, which should be about 130 to 140 degrees Fahrenheit. Turn the pile; water if necessary.

Day 7: Same as Day 4.

Day 10: Turn the pile and check the temperature, which should be cooling down to around 110 degrees Fahrenheit—the compost is almost ready.

Day 14: The compost should be ready for use.

I have found that my piles can heat up as high as 160 degrees Fahrenheit and that this method is reliable for killing any weed seeds that were in the raw materials.

Additional Advice

- For hot composting, a multiple-bin system is best so that you have somewhere to turn the compost into when rotating it. When the first bin has been turned into the second, the first is ready to be filled up again. Many people use a three-bin system; for me, four bins are ideal. By the final turning, the material is well finished and can be used almost immediately.
- Although the foregoing bin-building project uses fencing, any material that can form good, strong walls can be used, although there are pros and cons to each. Wooden shipping pallets, for example, provide good air circulation on all sides and can be readily found around Dumpsters in industrial areas, but they will rot after a few years, clog with compost, and provide a nice habitat for rodents. Cinder-block walls make a visually appealing containment system as well as a good solid structure. A solid multiple-bin system of stacked cinder blocks can be built for under $150 with no special tools or abilities. If you choose to, you can fashion doors from mesh fencing or wood (use your imagination) to go on the front openings; this will finish the bins with a tidy look while allowing the piles to be evenly stacked.
- For those with small or nonexistent yards, there are compost tumblers, which get rave reviews from everyone I know who has one. A tumbler is a rotating barrel that encloses the compost and aerates the materials as you rotate it. It keeps the compost away from rodents and other pests, blocks the sight and smell of decomposition, and minimizes the work of turning. It does limit the volume of your compost making, although this is not a problem for many urban dwellers, and it is still work to turn something as heavy as compost, no matter the device. It is possible to build one yourself, yielding varying degrees of success and function.

Keeping Chickens

Poultry keeping is one of the easiest and most fun hobbies that someone interested in food and gardening can have. The birds are hilarious, give you eggs so tasty that you'll come to dread store-bought ones, and will keep your compost pile hot and working well. They make great conversation (topics, although you'll find that they do tend to chat among themselves), and they don't shed on the couch. In many ways, they are ideal pets.

Why would you want to do this?
Chickens are wonderful pets, are great for kids, contribute the best compost additive for the garden, and are really fun to watch.

Why wouldn't you want to do this?
You don't want to care for animals, don't eat eggs, or don't want to deal with chicken manure.

Is there an easier way?
Buy a ready-made coop; there are many choices.

Cost comparison:
Giving your chickens kitchen scraps contributes to a reduction in their feed intake and a possible reduction in your garbage-hauling costs. Letting them roam your garden occasionally to eat bugs can reduce your need to buy pesticides.

Skills needed:
Concern and attention to animals and their needs. Basic construction skills for building your coop.

Learn more about it:
In addition to finding local or online sources, I recommend *Storey's Guide to Raising Chickens* (Storey, 3rd ed., 2010) by Gail Damerow, a fabulous all-around guide. *Poultry House Construction* (Domestic Fowl Research, 1997) by Michael Roberts is a go-to guide for building coops (the author includes plans for his designs). Wendy Bedwell-Wilson's *Starter Coops* (I-5 Press, 2012) offers illustrated coop instructions and care advice. *Keep Chickens!* (Storey, 2003) by Barbara Kilarski (from Portland!) is a lighthearted and confidence-inspiring read. *Chickens in Your Backyard* (Rodale, 1976) by Rick Luttmann and Gail Luttmann is an informative and fun basic book.

Raising chicks and then collecting the eggs when they are adults can be a fun and educational experience.

The coop must protect your birds from determined predators.

Several years ago, a friend told me that she wanted to get her husband some chicks for his birthday. Although she didn't want them, he did, and she decided that this gift would show him how much she loved him. He was thrilled to receive the chicks, and he and their boys commenced building the coop and pen.

He called me repeatedly with questions, so I sat down one evening and wrote what I called my "chicken brain dump" for him—every bit of chicken-related information I could think of at the time. I found out later that he passed it on to some of his friends, who have passed it on to some of their friends. It has become somewhat popular among local chicken-keeping beginners, and I'll share it with you here. When planning a coop and laying out a pen, keep in mind that for two to three birds, you'll need a bare minimum of 12 square feet in the pen, and a house that is no less than 4 square feet. Adjust your plans according to your own flock.

The Necessities

An appropriate chicken coop should

- be easy to clean out and stock with clean bedding;
- be strong and secure enough to keep out predators and rodents;
- protect your birds from drafts and wetness;
- have good ventilation (chickens are very susceptible to respiratory ailments);
- provide the birds with enough space to walk and stretch their wings and a place to roost;
- have some sort of natural light source.

A good pen or yard for the chickens should

- have good drainage;
- have a dry area where the birds can dust year-round;
- protect the birds with cover and shelter from wind, rain, and sun;
- have nests or protected areas in which to lay eggs;
- have sanitary and accessible food and water.

Note: Nests and food/water can be inside the coop if desired.

In their coop, chickens need the following items:

- A perch, set at a minimum of 4 to 6 inches off the floor. It is helpful to have a pull-out tray underneath the perch because this is where 90 percent of the droppings will go.
- 8 to 12 inches of perch space per bird. For two or

three birds, plan a roost bar of at least 24 inches; 36 inches is ideal. In hot and humid weather, the chickens will need a longer perch so they can keep more space between them.

- A nest box, although this can be outside the house. It is nice to be able to access the nest box from outside the pen so you don't have to walk in and get dirty when you collect eggs. Try to situate the chickens' entrance to the nest box so that they don't have to walk under the perches; otherwise, they will be walking through their droppings, which will make the eggs dirtier than usual. If the nest box is in the coop, place it lower than the perches—chickens are tree-perchers and ground-nesters.
- Bedding, which can be shavings bought at a pet store, sawdust, or shredded paper. Such materials help mitigate the waste, absorb the ammonia, facilitate cleaning, and balance out the extremely high nitrogen in the waste if you compost it. It also is comfortable for chickens to walk on.

It is helpful if you can close the door to the coop at night for warmth and to ward off nocturnal predators. However, if your pen is fully secure (five sides of chicken wire), then the door can stay open all the time. Birds deal better with cold (they wear down comforters!) than they do with heat.

Feeders are better outside the coop, under some type of cover to protect the feed from rain. When the feed is inside the house, it encourages the birds to stay in, which then encourages fighting, egg eating, messing in the nest, and broodiness (wanting to sit on the eggs to get them to hatch, which simultaneously stops the broody bird's ovulation). If you can hang the feeder, all the better, because chickens have a habit of scratching while they eat, which can get dirt into the feeder. A hanging feeder also avoids "billing out," which is the chickens' scooping feed onto the ground with their beaks and thus wasting feed. It is best to put out only as much food as will be eaten in one day; otherwise, rats and mice will begin to visit this convenient cafeteria.

It is also best to have waterers outside the coop in case of spills and leaks; they also will be easier to clean and monitor. An automatic waterer (gravity waterer) that hooks up to a hose or spigot is ideal, and the effort you save for the money you spend is well worth it. You can find these types of waterers at feed stores.

Types of Birds

There are three kinds of chickens—egg layers, "dual-purpose" birds, and meat birds. Egg layers are hens specifically bred for high egg production in their first two to three years. In the first year that a hen lays, she will lay smaller eggs more frequently (every day). In the second year, the laying becomes less frequent (once every one and a half to two days), but she will lay larger eggs. Often, this continues into the third year, but laying slowly decreases as the chicken ages.

Layers are smaller and leaner than other chickens, designed for egg production only. White Leghorns are typical egg birds—slim, slight, lean, mean egg machines. This stature also makes layers more able to fly than other chickens, at least up and over tall fences. When a layer breed bird dies or is culled, it is not good for eating because of its slim build.

Dual-purpose birds are exactly that—good for a couple years of egg production and then not bad for the stew pot. They are larger, meatier birds than egg layers, and they're good at laying too, often producing larger eggs in their second and third years. Rhode Island Reds, Barred Rocks, New Hampshire Reds, and Australorps are all technically dual-purpose birds—big birds that make a decent coq au vin when their laying time is up.

The Cornish Cross is a common meat bird.

The laying ability and larger build of the Rhode Island Red make it a good dual-purpose bird.

If you want meat, meat birds are a different type of chicken altogether. The most common is a hybrid called Cornish Cross, which is bred to produce large breasts and thighs, meaty and tender. They don't really roost and are kept only a short time, being fed to maximize their "fleshing out" and then processed at no later than ten to twelve weeks old.

Other hybrids are great in terms of not having many of the traits that can be annoying in chickens. For example, sex link hybrids have had the broody gene bred out of them, and although they are large enough to make decent eating when their laying time is over, they are extremely efficient egg layers—and are very friendly and docile as well. They come in black, red and gold; the Gold Sex Links, or Gold Stars, are my absolute favorite birds— beautiful, friendly, excellent layers. One of them was my first raccoon casualty, and one of two that I wept for.

Other Details

It's ideal to have three birds. It's unkind to keep just one, as chickens are flock birds, and two can make for a peculiar hierarchy.

When you buy birds, they can be *sexed*, that is, sorted and separated into *cockerels* (young males) and *pullets* (young females). *Straight-run* birds go directly from

hatching in the incubator into a brooder without being sexed and can be any mix of cockerels and pullets. If you are after egg layers, do not get straight-run birds. Pay the extra money to get assurance (90 percent with most hatcheries, higher if ordering sex links) that you'll have a hen. It's a disappointment to find that not only will you get no eggs out of your bird but also that he crows at first light and sometimes even all day. It's illegal in many cities to keep roosters (or to have more than a certain number of chickens without a permit).

From chick to laying is about five months, and it's not unusual to get double-yolked eggs when chickens first begin to lay (I once got a triple-yolker, and the world record is five yolks). Occasionally, there are shell-less eggs in the early days. Some breeds lay white eggs, while others lay brown eggs. Auracanas and their cousins, Ameraucanas, lay blue and blue to blue-green eggs, respectively.

Chickens will slow down and more often stop laying during months with fewer daylight hours; egg production drops significantly in the late fall and winter. There is a photoreceptive gland behind a hen's eyes that stimulates laying when triggered by a certain number of daylight hours. If you want your birds to lay regularly year-round, you can put artificial light on a timer in their coop, but this taxes their endocrine systems and, ultimately, their health. Hens are born with all of the eggs they are going to lay already in their bodies—you won't get any more than that by giving them extra light.

Chicken manure and bedding make great compost; they are fabulous activators for compost piles because they are very high in nitrogen. Don't apply manure or bedding directly to young plants because it will burn new growth, and too much nitrogen can delay blooming. However, nitrogen encourages lots of leaf growth, so these materials are good for herbs and salad when composted.

The chickens, if you let them, will wander around your garden, eating the pests and scratching in the soil. Their scratching aerates the soil and exposes weed seeds and bugs to the chickens' voracious appetites, and the chickens' manure fertilizes the area. They also eat young, fresh green plants, but they prefer to scratch in the soil and eat bugs. My hens eat all of my weeds and kitchen scraps, even the meat—they pick the bones clean. They will eat almost anything live that ventures into their pen and is smaller than they are—I've seen them catch and kill a mouse and then have a fabulous game of rugby with it. They can even track flies and snatch them out of the air.

You can use a dog run for a pen, and even a simple doghouse can be manipulated into a totally functional

Fresh eggs can be as close as your backyard, even in an urban or suburban neighborhood.

coop; either of these items can be found secondhand. Most chickens won't make much of an effort to escape if they are content, so you can use lightweight fencing, but you must shut them in at night to protect them from predators, such as raccoons, rats, coyotes, owls, and hawks, who will try hard to get in and kill them.

Build a Confined Coop

We know that chickens need a safe place to roost at night. The following is a design for a very basic enclosed pen that can be adjusted to fit your yard as needed. It's modular, so you can easily add, divide, or remove sections as your flock shrinks and expands. You can also modify this as a henhouse for a yard of free-range birds. Remember to include the previously listed chicken-coop essentials in your design. This project and the illustrations are shared with us courtesy of Wendy Bedwell-Wilson's *Starter Coops* (I-5 Press, 2012).

Materials:

- [ ] Two sheets of ½-inch construction-grade plywood, 48 inches by 96 inches
- [ ] Twelve 96-inch lengths of 1x4 pine
- [ ] One roll of 1-inch mesh poultry netting, 48 inches by 25 feet
- [ ] Approximately 500 1¼- to 1½-inch galvanized deck screws
- [ ] Tape measure
- [ ] Pencil
- [ ] Square
- [ ] Circular saw or handsaw
- [ ] Electric screwdriver
- [ ] Staple gun and box of staples
- [ ] Two 4-inch hinges
- [ ] 1 hook-and-eye safety latch

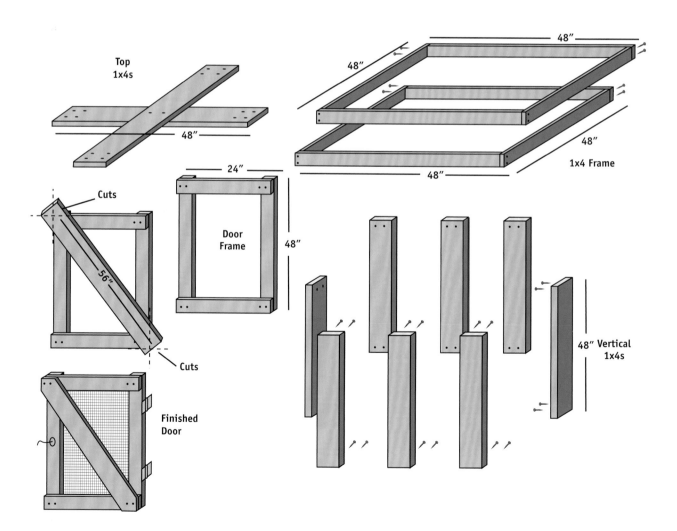

Step 1: Measure, mark, and cut eight pieces of 1x4 lumber, each 48 inches long. Screw four of the pieces together to form a square (alternate inside and outside boards), using two screws at each joint to ensure a solid connection. Then screw the other four pieces together to make an identical square. You now have the top and bottom of your frame.

Step 2: Measure, mark, and cut eight more 48-inch pieces of 1x4 lumber to create the vertical wall pieces. Use one at each corner and one in the center of each side (at the 24-inch mark) to attach your top and bottom frames, using two screws at each joint.

Step 3: Cut two more 48-inch lengths of 1x4 pine. Screw these 1x4s to the tops of the four center vertical wall pieces so that they cross the top of the coop. You should now have a cube-like frame that measures 48 inches on all sides.

Step 4: When the structure is completely framed, enclose the left half of the front, the left half of the top, and the front half of each side with poultry wire. (The door will occupy the other half of the front, and plywood will cover all other open areas to prevent drafts. The mesh areas allow air and sunshine to flow into the coop.) Measure and cut enough wire for each panel to cover the panel with excess for staples. Affix the wire inside your frame using your staple gun, placing a staple about every 2 inches to prevent predators from breaking in.

Step 5: Measure, mark, and cut three pieces of plywood for the sides and roof, each 24 inches by 48 inches, and one piece for the back that is 48 inches by 48 inches. To attach the plywood to the open sections of the coop,

place one screw at each corner and screws about every 6 inches in between.

Step 6: To build the door, first, cut five pieces of 1x4: two 48-inch pieces and two 24-inch pieces for the frame, plus one 56-inch piece that will act as a cross-brace. Next, lay the 24-inch pieces on top of the 48-inch pieces to create a rectangle and screw them together using four 1½-inch screws at each junction. Then, measure and cut a 24-inch-by-48-inch piece of poultry wire and staple it to the back of the door. Screw in your cross-brace from the top left to the bottom right, using two screws at each end. Trim the edges of the brace with your handsaw to match the angle of the door's frame. Finally, screw your hinges to the door's frame, attach the door to your coop's frame, and add your safety latch.

Build Nesting Boxes

You can expand this simple double nesting box to as many boxes as you need, as well as include hatches at the top or back for easy egg-collecting. Each box should measure about 12 inches wide by 14 inches high by 12 inches deep for an average layer. Project and illustrations from Wendy Bedwell-Wilson's *Starter Coops* (I-5 Press, 2012).

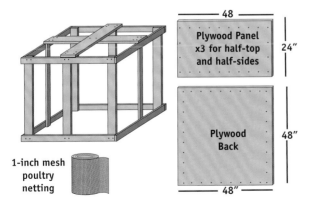

1-inch mesh poultry netting

48 — Plywood Panel x3 for half-top and half-sides — 24"

Plywood Back — 48" / 48"

Materials:

- [ ] One sheet ¾-inch construction-grade plywood, 48 inches by 96 inches
- [ ] One 22½-inch length of 1x2 pine
- [ ] Two 12-inch lengths of 1x2 pine
- [ ] One 24-inch length of 2x2 pine
- [ ] Approximately fifty 1½-inch galvanized deck screws
- [ ] Electric screwdriver
- [ ] Tape measure
- [ ] Pencil
- [ ] Circular saw or handsaw
- [ ] Electric drill

Step 1: For the base and top of your nesting box, measure and mark on the plywood two pieces that are each 24 inches by 12 inches. Carefully use your circular saw or handsaw to make the cuts.

Step 2: For the sides and center divider of your box, measure, mark, and cut two pieces of plywood that are each 14 inches by 12 inches (sides) and one piece that is 12½ inches by 11¼ inches (center divider). Set the center divider aside for now.

Step 3: Screw a side to each end of the box's base, placing one screw at each corner at least 1 inch from the edge and one screw in the middle of each side.

Step 4: Use your screwdriver to attach the top to the box's sides, using three screws for each side.

Step 5: For the back, measure, mark, and cut a piece of plywood that is 24 inches by 14 inches. Secure the piece in place with one screw about 2 inches from each corner and another about every 5 inches in between.

Step 6: Lay the nest box on its back and slide the pre-cut center divider piece directly into the center of the box so that it fits snugly against all sides. You should now have two boxes of equal size. Secure the divider in place by driving two screws each into the top and bottom and three screws into the back.

Step 7: With the box still on its back, add a lip to the front of the box by attaching the 22½-inch length of 1x2 pine across the bottom front of the structure. Drive two screws through each of the box's sides and into the lip and one through the lip and into the base about every 6 inches in between, making sure the screws drive securely into the plywood and don't poke out at all.

Step 8: Add a 12-inch length of 1x2 pine to the bottom of each side of the box, allowing 8 inches to extend out in front of the box. Drive two screws through each of these arms into the sides of the box. Connect the two arms with the 24-inch length of 2x2 pine that will serve as a perch. Securing the perch by driving two screws through the end of each arm into the ends of the 2x2.

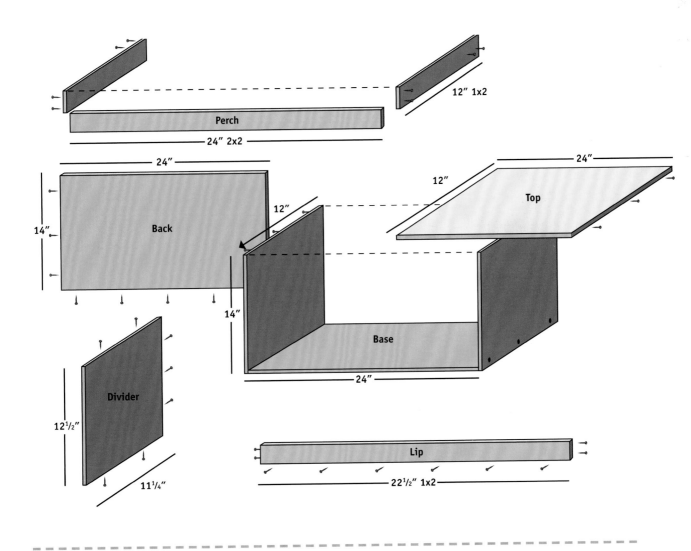

12" 1x2
Perch
24" 2x2
24"
Back
14"
12"
Top
24"
12"
14"
Base
24"
14"
Divider
12½"
11¼"
Lip
22½" 1x2

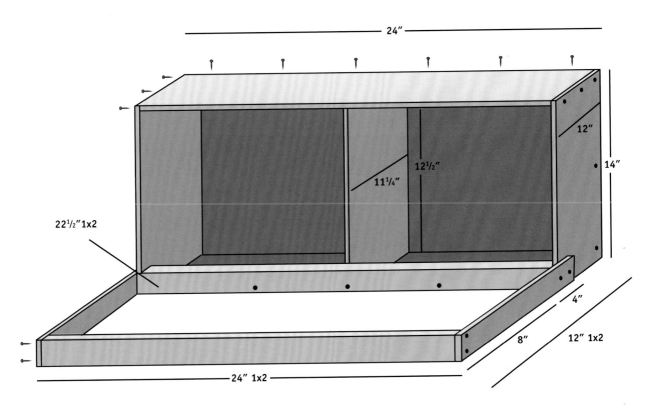

24"
12"
14"
12½"
11¼"
22½"1x2
4"
8"
12" 1x2
24" 1x2

Dairy in the Backyard: Goats

While most commercial dairy products come from cows, you can get fresh dairy right at home, even if you have a small yard or small plot of land. Goats are great animals for urban and suburban farmers. I can extol the virtues of goats all day!

Why would you want to do this?

You're interested in fresh milk, healthy home-made dairy products, an endless supply of mulch and compost makings for your garden, an in-yard garbage disposal for extra yard trimmings, and a friendly, generally quiet, companion animal.

Why wouldn't you want to do this?

You don't have enough space; you aren't able to be consistent about milking once or twice daily; you don't mind buying your milk; or you have a fear of raw dairy products.

Skills needed:

Keeping goats falls under the category of owning and training an animal and the associated responsibilities, much the same as owning a dog. You'll need time each day and an affinity for animals. Once you get into the swing of things, you'll be amazed at how easy it can be and how much fun it is.

Learn more about it:

Storey's Guide to Raising Dairy Goats (Storey, 2001) by Jerry Belanger is a fabulous starter book to get you thinking about and planning for goats with accurate information; I have referred to it countless times. *Raising Goats for Dummies* (Wiley, 2010) by Cheryl K. Smith is the best book I've found on the topic (although I dislike anyone being called or calling him- or herself dumb). I have since met the author and can attest to her knowledge and experience with all things goat. There is nothing that I have experienced in the caprine world that this book or Cheryl herself cannot address. It truly is my bible for goat owning.

The goat-keeping process is far more involved than this book will address. I highly recommend reading the aforementioned reference books if you want to get serious about owning goats. To help you consider a possible future with goats, let me start by clearing up some misconceptions.

Common Misconceptions

1. *Goats are large animals.* Goats can be big. However, like dogs, goats can range greatly in size (and temperament), having been bred for various purposes. I have a neighbor who has three full-size goats, and their shoulders reach the bottom of my rib cage (I am 6'1", so that's around 4 feet—at their shoulders). That is fairly big. None of the three is a milker, but when a full-size goat does give milk, you are looking at upward of a gallon a day. A gallon a day is a lot of milk.

As the demand for smaller goats for smaller acreage has increased, the number of breeds and varieties of smaller goats has increased. Pygmy goats are tiny things,

Milking Through

A note about milking: dairy animals give the most milk shortly after delivering their babies, and milk production will drop off over time. Many goat owners milk their dams (mothers) for only ten months and then rebreed them to obtain a high output after kidding (giving birth) the following year. This is not necessary. There is a practice called *milking through*, which means that the owner continues to milk the dam as long as she will give milk. Success depends on many factors, including the genetic makeup of the goat (a good milker from a good milking line), your patience and tolerance of a varying milk yield throughout the seasons, and a solid routine, followed religiously. My first doe had been milked for more than two years straight when I got her, and I milked her for two and a half years more before "freshening" her (stopping milking ["drying off"] and rebreeding).

not much larger than medium-size dogs, weighing from 20 to maybe 60 pounds maximum. They look like miniature versions of full-size goats, are friendly and easygoing, and can be quite prolific milkers. A couple of years ago, I boarded two pygmies—a brother and sister—whose owner was between residences and didn't have a yard. When she came to visit them and take them for walks with their collars and leashes two or three times a week, they got lots of attention from onlookers.

There is also a small breed called the Nigerian Dwarf. These goats are physically different from full-size and pygmy goats in that they have shorter legs and more barrel-shaped bodies with big bellies. They have wonderful, friendly temperaments and are reported to give very sweet, tasty milk. The Nigerian Dwarf doe I am most familiar with is owned by my friend Sheri. The doe gave eight cups a day right after delivering her four kids, which was her fourth pregnancy; her milk production tapered to two cups at the end of her milking days.

The size, personality, and milk output of the Nigerian breed was perfect—and remains so—for Sheri and her backyard. Breeders who felt that this breed was too small began to crossbreed Nigerian goats with goats of full-size breeds, creating what are known as *minis*. My milker is a mini LaMancha doe that is taller than her Nigerian Dwarf mother and shorter than her full-size LaMancha father. She weighs less (90 pounds) than our sheepdog (103 pounds) and has shoulders that come to my hip. She rides in the back of my station wagon to the vet, to other farms for breeding, or down to Sheri's house for boarding and milking when I go on vacation. She actually enjoys riding in the car, and I can't count the number of people in other cars who point, honk, stare, and take photos when they notice her!

After having three kids almost a year ago, my mini LaMancha is currently giving a full quart of milk daily, down from almost a half gallon each day in the spring and summer. I use 1 to 2 gallons of milk each time I make cheese, so I try to make cheese or yogurt weekly because the milk supply is steady.

2. *Goat's milk tastes bad.* For years, I bought goat's milk from the store, and at first I referred to the taste as "goaty." In time, however, I began to realize that the taste was sort of "hay-ey." Once I got my own goats and drank their milk, there was none of that taste for me (although my son would disagree). I have compared the

tastes of milk from many different goats with commercial goat milk, and I've come to my own conclusions. My goat gives me slightly sweet milk with a rich, creamy taste that is fresh and almost mildly flowery. I love it.

Goat's milk will vary in taste depending on the goat, and milk that comes from farms with bucks in residence tastes much more "goaty." Bucks have a natural musky odor that emanates from their glands and permeates everything that they come in contact with (or near, in some cases). I personally believe that it can seep into the milk of a lactating doe that lives with an intact buck. Neither Sheri nor I have ever had that buck smell or taste in our milk from four different milking does.

3. *Goats smell bad.* As with many generalizations, this is relative. Depending on your own sense of smell, yes, goats can smell bad. Bear in mind, though, that to some people, so can onions, garlic, fresh cilantro, beer brewing (this affronts my husband), or even chocolate (my own peculiarity).

Buck smell can be considered offensive—Sheri's daughter Elyssa can barely breathe when there is buck scent around. I, on the other hand, love the smell, and I loved to pet and scratch the buck that lived with us for a couple of months. The thick musky smell is determined by hormones, which are strong and constant in these guys.

A barn that has only female goats or wethers (castrated male goats) smells like, well, a barn, but the bedding should absorb the odor almost immediately. If the barn or pen smells bad, something is wrong. It may be that you are not cleaning often enough or one of your charges is sick and needs treatment. A freshly cleaned barn and healthy goats smell sweet and livestocky, which to me is a nice smell (much better than wet dog!).

4. *Goats are loud.* This can be true, but it depends on the goats. Although I have been lucky enough to have mostly quiet does and tolerant neighbors, too much noise can be a dealbreaker when a goat pen borders neighboring properties. Goats, like chickens, are not nocturnal, and they shouldn't make noise at night unless there is a problem. Certain goats, though, can be loud and will make disturbing noises during the day. Sheri gave me her loud goat, which happens to be my milker's daughter, and she isn't as bad at my place as she was at Sheri's. I believe that some of the noisemaking can be situational.

Nigerian Dwarf goats are small and friendly.

5. *Goats can be unmanageable.* You can control—or cause—a goat's behavior. The most important part of goat keeping is to get goats (never fewer than two, as they are herd animals) that you can physically handle. Herd or flock animals have an innate hierarchy, and there will always be a herd queen (or leader if you have both wethers and does). The herd queen will constantly challenge you until you've established your dominance.

If your goats are bigger than you, or if they have horns, or if you have children who are not able to stand their ground with the goats, then, yes, goats can be unmanageable. Goats do best with someone who can be consistent, strong, loving, and kind. They are very smart animals, and they can be trained (some goats will leave the barn to urinate outside), but their wits need to be bested. If they understand that you are in charge of them and will take care of their welfare, they can establish their own hierarchy within the herd and live happily together, with you as their overseer.

6. *Goats eat everything.* This is a load of hooey and one of the biggest misconceptions about goats. Goats explore with their mouths, and personable goats may seem to want to chew everything, especially hanging things. They nibble on projections from clothing, such as tags or buttons, to try to get your attention and get you to pet them. Goats actually have very sensitive digestive systems, and although they will eat thorny things (e.g., blackberries, holly, thistles) with nary a second thought, there are long lists of things that they should not have

Goats are curious and are talented climbers.

access to, such as rhododendrons, azaleas, camellias, and certain other plants. I have noticed that my goats seem to sense that some plants are bad for them—they avoid certain plants that would poison them if ingested—and they are also very good at finding browse as they need it (for example, picking out weeds high in a mineral that is lacking in their diets). Goats do not eat tin cans, but they may appear to when they go after the labels, which they like because paper is a wood-pulp product.

Variations in Flavor

If goat's milk has a strong taste, and the doe does not live with a buck, there can be other reasons. Some owners say that the milk's flavor can depend on what the goat eats; for example, onions, garlic, or some strong-flavored plants are said to impart flavors. I've never found this to be the case, but then my goats won't eat onions or garlic. Sometimes, off-flavored milk can indicate an illness, such as mastitis, or environmental pollutants, which you should pay attention to.

Are you still thinking about owning goats? I hope so. Now it's time to look at some of the more practical considerations.

Practical Considerations

1. *Does your community allow you to keep two or more goats?* Most cities count goats as livestock and therefore have limits or regulations regarding their keeping; some municipalities do not allow them at all. If your community does not allow goats, don't push it. My goats are pets, even though I have a livestock permit for them. Like any pets, goats should not be acclimated to a living situation unless they will spend most or all of their lives there. Change is very stressful to goats.

2. *Do you have space for goats?* A herd of goats (two or more) needs a secure, sheltered area in which to sleep and find protection from the elements, as well a securely fenced open area in which to roam safely. Depending on the size of the goats, their accommodations can range from a dog house and a large dog run to a small barn or a converted shed and a large fenced field. You don't even have to keep them in your backyard. Portland has several herds of goats that belong to farmers but live on school campuses and vacant city lots to assist in keeping down the weeds and shrubby overgrowth.

Cheryl Smith gives the following advice in *Raising Goats for Dummies* on how much space you'll need:

If you live in an area where you can't let your goats roam over a large area, you need about 20 square feet per adult standard-sized goat for sleeping and resting, plus another 30 square feet (outdoors, ideally) for exercise. If you have a larger outdoor area in which to raise your goats—where they'll have pasture, woods, or range—you need less indoor space per goat because they'll only rest and sleep there. The rule of thumb is 10 to 15 square feet per adult standard-sized goat.

We built an 8-by-15-foot "barn" that has one-third of it closed off for milking and storage. The remaining 80 square feet easily houses four minis, but I have had more in the past. Their fenced yard is ample, about 3,500 square feet at forest edge. Sheri's Nigerians live in a converted shed of 12 by 16 feet but have only slightly more than 500 square feet of fenced yard.

3. *Do you have or can you afford a stable shelter and strong fencing for your goats?* When I fenced my pen in preparation for goats, I was tuned into chickens, and I made sure that the fence was very secure at the bottom (thinking of predators that could dig under). Little did I know that the two most important factors for goat fencing are height (minimum of 5 feet) and strength (they will lean and rub against the posts and fence to scratch and will stand up against the fencing to explore). Goats, like deer, are browsers, eating bark, branches, and leaves on trees. They prefer to jump, climb, and explore upward areas (remember that they explore with their mouths).

4. *Do you want to milk your goats?* If you plan to milk, you have a number of additional considerations. How will you get a goat that gives milk? Will you breed her? Will you buy one that is pregnant, and, if so, are you willing to deliver the kids? Or will you buy a doe with (or without) her kids? Do you have space to milk, separate from the rest of the herd? Are you willing to build or buy a stanchion?

More importantly, are you willing and able to provide the additional high-quality feed that a lactating doe requires? Are you willing and able to devote regular time at least once a day (better yet, twice) to milking your doe? Every day? On weekends and holidays? If you go on vacation, do you have someone who can either take the goats or come and milk them for you?

All of these are extremely important questions if you want to have a steady supply of goat milk. I'll be honest: it's not easy, especially when the novelty wears off. It has changed my life in certain ways (fewer vacations and having to schedule appointments so they don't conflict with milking times), but I have no regrets and have found that I really love goat keeping, even with the effort that it entails. You'll find your balance also.

5. *Are you willing to commit to keeping and caring for goats for their entire lives?* You should plan to have your goats for at least as long as you would have a dog— a healthy goat can live for eight to fifteen years and thrives on stability and a connection with its owner and its owner's people. If you buy a goat and then decide you don't want it, you'll have a hard time rehoming it. Our local humane society will not take goats, and not all

Do you have room and time for a herd?

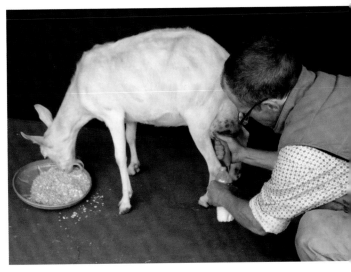

Milking requires time and consistency on your part.

areas have farm-animal-rescue facilities. It is possible to sell your goats, and in the case of too many kids or a herd that is getting too big, this is a good plan of action. However, a change in living situation is extremely stressful and unfair to a mature goat because it wants to bond with its owners. If you can't make a long-term commitment to goats, don't get them.

If I haven't scared you off, check out the books that I mentioned on the first page of this project. Owning goats is one of the largest endeavors I have embarked on, and it's most definitely my favorite so far.

Mason Bees

Keeping bees is a great pursuit for the urbanite, and developing a population of orchard mason bees is an easy and fun way to introduce yourself to the invertebrate pollinator's world. Native pollinators, after all, contribute much to our food system and the pollination of any garden.

Why would you want to do this?
If you are interested in bees and are considering honey bees, try mason bees first and take the time to learn more about honey bees.

Why wouldn't you want to do this?
You have a phobia of bees or an aversion to insects.

Is there an easier way?
You can buy premade mason bee homes, which are usually wooden blocks with predrilled holes and paper straws to insert into the holes. They are not too expensive and, depending on where you buy them, part of the purchase price may be donated to the understanding, study, and protection of native bee populations.

Cost comparison:
The bees will cost more than the materials for the house; your homemade house will be less expensive than a premade one.

Learn more about it:
Pollination with Mason Bees (Beediverse, 2002) by Dr. Margriet Dogterom is good beginner's book by a specialist on orchard mason bees; *Crown Bees* (www.crownbees.com) calls itself "Your Complete Mason Bee Resource" and offers basic information and products to new beekeepers; *Mason Bees for the Backyard Gardener* (Inkwater Press, 2010) by Sherian A. Wright includes step-by-step instructions and color photographs for creating habitats.

Mason bees are hard-working pollinators.

Despite my mother's keeping honey bees when I was growing up, I came to beekeeping only in the last couple of years, and I'm still surprised that it took me so long. The bees are a natural addition to any garden, and beekeeping requires far less maintenance and less money than keeping chickens or goats does. Everyone interested in urban farming (except those with a severe allergy to bee stings) should acquaint themselves with honey bees. Not too long ago, I was diagnosed with a moderate ("large local") allergy to bee stings (most people have an "ordinary local" reaction) and was warned that it could escalate to a severe allergy at any time. I was discouraged, but I decided that my garden, my interest, and the environment needed me to continue to keep my hives. I stock up on epinephrine injectors, keep prednisone at hand, and attempt to stay calm when working my hives. The bees are worth the risk.

I like to advocate keeping honey bees to everyone who's interested in gardens. We've all heard about the decline in the honey bee population, and it is said that one in every three bites of the food we eat comes from bee pollination. If the numbers of bees diminish, our ability to have the variety and amount of produce to which we are accustomed takes a hit too. Bees are wonderful creatures, and there are very few reasons not to pursue beekeeping for people who are interested in it.

If you have an interest in keeping honey bees, two helpful books to start off with are *Beekeeping for Dummies* (Wiley, 2009) by Howland Blackiston and

Honey Bee Hobbyist by Norman Gary, PhD (I-5 Press, 2010). I also advise finding a local beekeeping organization—check with the American Beekeeping Federation (www.abfnet.org) for a chapter in your area.

Also get to know a beekeeper in your area. Ask questions and spend some time with him or her, watching and learning. Get a feel for the protective garb in different types of weather, and experience the sound and feel of thousands of honey bees buzzing around (and on) you. How do you feel being around a hive, with the possibility of being stung? Keeping honey bees is not something to do on a lark, so think about it carefully. If, after considering all aspects, you still want to try it, then go ahead.

Along with a reduction in honey bees, there has also been a dramatic reduction in the populations of native pollinators, especially as developed areas grow at the expense of wildlands. Keeping one of these pollinator species lets you dip a toe into beekeeping in a very simple manner. You won't get honey from these bees, but they still contribute to pollination and your beekeeping education.

Mason bees are big, fat, black bees who are very industrious and practically never sting (the males don't have stingers and the females use theirs predominantly as *ovipositors*, or egg guides). Their main objective is to collect and "package" pollen and nectar, which pollinates the visited flowers. They place each food package in a tubular space (or straw), lay a single egg on top of

Look for evidence that mason bees have visited the "hive."

it, and then seal the chamber with mud. This process is repeated along the line until the selected spaces are filled with egg chambers for next year's hatch. Their prolific egg-laying also multiplies their numbers exponentially. The process of sheltering them and fostering a community is so simple that, if you're interested in beekeeping, there's really no reason not to do it.

Building a Mason Bee House

Materials:

- ☐ Sheets of plain white 8½-by-11-inch paper cut either in half (two 8½-by-5½ inch pieces) or in fourths (four 8½-by-2¾-inch pieces), depending on how big you want the tubes
- ☐ Scotch tape
- ☐ New, unsharpened pencil or a piece of $5/16$-inch dowel pencil length or longer
- ☐ Piece of 3-inch or 4-inch PVC or ABS pipe, at least 9 inches long, with a cap on one end

It's fascinating to watch the mason bees at work.

Step 1: Lay the pencil or dowel along the 8½-inch length of one of the pieces of cut paper.

Step 2: Tightly roll the paper around the pencil (keeping the metal top outside the roll so it's easy to remove later) or dowel.

Step 3: Tape the edge of the paper securely to keep it in a roll.

Step 4: Slide the pencil or dowel out of the tube.

Step 5: Even out the edges of the tube as much as possible (don't spend a lot of time on this).

Step 6: Pinch and fold over one end of the paper tube.

Step 7: Lay the paper tube inside the pipe with the fold at the capped end of the pipe.

Step 8: Repeat the process until the plastic pipe is full of paper tubes.

You can hang your mason bee house anywhere outside in an east- to southeast-facing direction, preferably under cover, such as a porch or roof overhang. Watch as the weather gets warmer to see if native bees in your area find it. They start coming out of their nesting holes in the spring as temperatures rise to consistently above 55 degrees Fahrenheit, so by mid- to late spring, depending on where you live, you should see some mud-filled straws.

Native bees may not find your bee house immediately, so if you buy a few straws of mason bees and tuck them, mud-side out, in the midst of your homemade straws, you'll most likely see your straws filling up with little mud nests by mid-summer. In early fall, I bring my mason bee houses into my shed to protect them from foraging woodpeckers (insectivores love the straws and will happily gobble them up) and from the deep freezes of winter, hanging them back outside in late March for the process to start again. In temperate winter climates, you can attach a piece of screen around the open end of the pipe with rubber bands, but remember to take it off in the spring before the bees start to hatch. If you've got a lot of filled straws, you can make more empties and put them, along with two or three filled ones, into a few new houses for multiple "hives" around your outdoor area.

SECTION IV:
Outdoor Gardening

Create a Garden Area

I f you have a yard, no matter what size, you have a place to grow food. There are many projects out there that focus on different ways to garden depending on the shape of your space, but this simple project turns just a square or rectangular piece of your yard into a growing area.

Why would you want to do this?

You want to try growing your own food and to taste the difference of homegrown produce, such as tomatoes, peas, beans, and carrots. You can always turn the garden back into lawn if it doesn't work out.

Why wouldn't you want to do this?

You don't have any unpaved outdoor space, or you just don't want to grow your own food.

Is there an easier way?

Sometimes it is easier to hire someone to do the labor for you, especially if you have any physical limitations, but if you have the capability, it can be more rewarding when you do it yourself. If you dig your own garden area, you may even be able to skip a few sessions at the gym!

Cost comparison:

Once your garden is up and running, your grocery-store-produce costs will diminish greatly.

Skills needed:

A basic willingness to work hard and be patient with small successes over time.

Learn more about it:

Two of my favorite resources that I return to year after year are *The Essential Kitchen Gardener* (Holt, 1991) by Frieda Arkin and *How to Grow More Vegetables* (Ten Speed Press, 8th ed., 2012) by John Jeavons.

The aching back and dirty hands will be well worth it when you're enjoying your homegrown fresh produce.

Whenever I read an article about starting a garden, I find that the first piece of advice is to pay close attention to site location. Soil and exposure—to light, wind, and the elements—are the two most important factors. It is possible to slightly alter the exposure of your growing area once it is established, whether by trimming trees selectively to increase sunlight, putting up shade cloth and screens to reduce sunlight, or planting hedges to reduce wind and increase warm microclimates. To a certain degree, however, you are stuck with the land that you have. Soil, if poor, can be worked and amended for improvement, but the ultimate consistency and pH of the native soil are fixed factors.

If you want to create an outdoor garden area, look for a spot with optimal exposure and soil quality; ideally, you'll be able to find one site that has both. If this is not the case, you'll have to decide which is more important to you and consider if the other factor can be achieved with some modifications.

Consider the following when selecting a garden site (if you don't have a suitable piece of ground, see the chapters on raised beds, planters, or community gardening for other ways to start a growing area):

• *Does the site get good sun for at least six hours a day in spring and fall?* If not, could it get more sun with selective tree-trimming or structure removal? If you want to grow warm-season vegetables and fruits (e.g., tomatoes, basil, eggplant, peppers, squash, beans, melons), you'll need a bare minimum of six hours of full sunshine a day and ideally southern or western exposure. If you don't mind limiting your growing to cool-weather crops (e.g., cabbage, broccoli, lettuce, kale, peas, Asian greens), you can squeak by with less (but not much less) sunlight. The cool-season crops still need the light, but they can handle indirect light and partial shade whereas the warm-season crops really need full sunlight and heat to flourish.

• *Does the site have good soil?* Are there always some puddles, even in drier times? Are there a lot of weeds? Does nothing grow there, even during the wet seasons? Is there access on at least two sides for you to work the ground? Is the site up against a fence or a wall? Is there space around it to walk or push a wheelbarrow?

You can always amend soil. Double-digging and adding copious amounts of compost and manure can make even the worst sites fertile. Long-standing puddles indicate either overly compacted soil or drainage problems; barren soil with nothing growing can indicate serious nutrient deficiency. The presence of weeds can be a good sign, and the types of weeds can help you ascertain details about the chemical makeup

You don't need a large area to create a place to grow food.

Tending the garden is something the whole family can be part of.

of the soil. You'll need to have at least two sides of the space open to access your plants, and you'll need enough space to be able to walk and transport amendments and tools. An adjoining fence or wall can be good for trellising or climbing plants, but it can be problematic if it casts shade or if it reflects excess heat in hot seasons.

- *Is it in an area where animals might deposit waste or where unaware pedestrians might walk on it?*
- *Is it near any large trees with roots that would impede smaller plants?*
- *Does it have access to water?*
- *Is there enough space for you to plant what you want to grow?*
- *Is the site located conveniently for you to see and tend?*

Once you've found the spot that will become your garden, decide what you want to grow and how much space you want your garden to occupy as a whole. It's helpful, but not necessary, if you can calculate how many of each plant you will need per person in your family and the approximate square footage that those plants will require.

Next, figure out how you want to water your garden. If you routinely go outside for a bit each day, do you want to rely on a spigot and watering can? Watering everything by hand is by far the most beneficial for the garden and plants—the best thing you can do for your garden is give it regular attention to monitor its progress and problems—but it is also the most time- and labor-intensive method. You can compromise by watering by hand with a hose that extends to the garden area, as well as having a sprinkler that you can hook the hose up to as needed. This allows you to hand-water most of the time but use a sprinkler when the garden needs a long, deep drink.

A single sprinkler for the whole garden may not be the best answer, especially if it is a large or irregular area. Another factor to consider is that some plants need lots of water, while others not only tolerate but actually thrive on less (e.g., tomatoes develop better flavor if you cut back on watering as the fruit begins to color). So giving all of the plants the same amount of water won't bring out their best. One more point to bear in mind: a dedicated watering system for the garden is great, but it allows you to be less attentive to the garden, which means you may not always notice problems or pests in time to prevent damage. That said, if you want to install a subterranean system, do it now before placing the garden.

Once you've decided on your area and have all of your plans in mind, you can really get started.

Materials:
- ☐ Shovel
- ☐ Long hose, long string, or white flour
- ☐ Strong back
- ☐ Optional: Pitchfork or pickax

Step 1: Stand in your garden-to-be and draw out lines that mark its outside shape and dimensions. Lay out the lines with hose or string, or sprinkle lines with flour. All three of these materials can delineate nice curves; for straight lines, you can use long boards to help you keep the edges straight.

Step 2: Start digging! A straight-edge shovel is best for edges, but use whatever you've got. Cut the edge lines along your drawn border, and turn every shovelful of soil onto itself. Do this for the whole area, flipping each shovelful upside down. You can chop the soil as you go,

or you can do the digging and turning with a pitchfork (or pickax) to really churn up the soil. The purpose is to loosen and mix up the soil as much as possible.

As you dig, remove any rocks that are bigger than your fist; smaller rocks can stay. Remove any weeds, shake the soil off their roots, and throw them out. Note whether you see many or few worms, and watch for bugs such as spiders and beetles—maybe set them aside for further inspection if you don't recognize them and you're not squeamish.

Step 3: Once you've chopped up the whole area and turned everything as inside out as possible, test your soil. All of the experts tell us to test our soil, and I've ignored this advice in the past, to my regret. Remove your blistered hands from your shovel and test your soil, right now, right here (after you crack open a beverage of choice and congratulate yourself on your hard work).

At this point, you ought to have a fair sense of what kind of soil you have. If it is thick, heavy, and difficult to break up, it is probably clay (which can contain lots of nutrients). If it is loose and dry and won't stay in a pile or falls out between the pitchfork tines, it may contain a lot of sand (which can make for great drainage). Lucky you if it is dark, it is somewhat moist, and it breaks up easily but holds in a clod when squeezed—you have loam! Chances are, you'll have a mixture of these three soil types.

Digging allows you to delineate your garden area, remove weeds, and prepare your soil for amendments, if needed.

Even if you already know what type of soil your area is reputed to have, and you recognize it as you dig, it's still worth it to test your soil. Not only will you give your back a chance to recover from all of the digging, but by stopping to test your soil, you'll understand what you'll be facing with your garden. Because soil types and testing methods vary widely from region to region and town to town, consult your local county extension department for soil-testing sources and how to decipher the test results. Soil tests run the gamut from DIY kits available at your local garden center to tests that require you to send samples to a lab, and the results will tell you what sort of amendments to add to your soil to bring it as close to a neutral pH (7.0) as possible. You will probably also learn about the structure of your soil (clay, loam, sand), how your soil type will affect drainage and plant growth, and what types of plants will fare best in your site.

Step 4: If you will be doing raised beds, now is the time to lay them out (see Project 7 in this section). If you are going to garden directly in the ground, skip this step.

Step 5: Mulch. You can lay down sheets of cardboard (a single layer will suffice) or a layer of newspaper (ten to twenty-five sheets thick) and top it with grass clippings, kitchen waste (no meat), autumn leaves, conifer needles, farm-animal manure, purchased compost, bagged soil, or a truckload of topsoil.

Planting Advice

By digging up and turning everything, you not only have loosened the area for roots but also have upturned long-buried weed seeds that will be so happy to be getting air and light that new weeds will jump out of the ground. However, when you cover the ground with biodegradable mulch ASAP, your plant roots will be able to grow down through it, but the weed seeds won't have a chance to germinate. The deeper you pile the mulch, the richer the soil will ultimately become from natural composting and worm activity. You can incorporate into the mulch any necessary soil amendments recommended by the soil test (you did do a soil test, right?). Water the mulch well and let it settle a bit.

If it is the fall, pile the mulch as deeply as possible and leave it for the winter. In the spring, when you start to

Mulch your garden liberally before planting.

All gardens benefit from hand-watering and daily attention.

plant, you'll find a thriving population of soil creatures in the lower area, and the roots of your plants will happily dive into this rich substrate. If it is the spring, make the final topping on your garden one of soil or compost. You can plant seeds directly in the top layer or dig a hole through it and set transplants in the hole.

Keep an eye on everything you plant, keeping seeds well moistened until germinated and well established, and watering transplants well to make sure they have good root-to-soil contact and grow long enough to find their own soil access. Enjoy the literal fruits of your labor.

Season Extension

Thanks to modern developments, it is easier than ever to make the growing season longer than nature dictates. For those who live in borderline temperate climates, season extension makes it possible to maximize the potential of their gardens; for those who live in areas with more extreme weather, it enables them to grow and harvest foods that would otherwise be unobtainable in their gardens.

Why would you want to do this?
You enjoy harvesting fresh food and would like to elongate the period of harvest, or you want to get a head start on the planting season.

Why wouldn't you want to do this?
You don't have much of a garden area, or you don't need cold-weather protection for your plants.

How does this differ from a store-bought version?
You can customize the following DIY cold-frame project to your own garden space.

Is there an easier way?
Manufactured cold frames usually cost more than the materials to build your own; however, sometimes the purchased versions are easier to manage. But where's the adventure in that?

Cost comparison:
You can make many season extenders cheaply by reusing things you already have around the house or finding salvaged materials for low to no cost.

Skills needed:
Basic construction skills.

Learn more about it:
Gardening Under Cover (Sasquatch Books, 1989) by

William Head; *Four-Season Harvest* (Chelsea Green, 1999) by Eliot Coleman and Barbara Damrosch.

The familiar greenhouse.

A hoop house is large enough for a gardener to enter.

Even though where I live is considered a "temperate climate," we can still have some extremes in temperatures and weather conditions. When I prepare my soil well and protect it and my plants from the elements, my growing season is longer, beginning earlier and extending to later final harvests.

Crop protection is nothing new. In 50 BC, the Roman poet Seneca described using mica in structures to surround and protect growing plants. Glasshouse use in Europe was first recorded in the late 1400s, with Charles VIII of France bringing back the idea of an *orangerie* from Italy, where he saw citrus growing in a protected environment. Today, we have polymers and plastics that can withstand elemental assaults and ultraviolet degradation over time, increasing our ability to design structures of almost any shape and size.

There are various styles of plant protection. Loose protection laid over a row of plants is called a *row cover*. Long semicircular tube structures over a bed are called *hoops* or *tunnels*; tall versions that people can walk through are *hoop houses* or *high tunnels*. Most everyone is familiar with *greenhouses* (the house-shaped structures also called *hothouses*, *conservatories*, and *orangeries*), but not everyone knows that they can be constructed of wood, metal, or fiberglass and that the clear material (called *glazing*) can be either glass or plastic.

Smaller bottomless structures over beds or soil are called *cold frames*; they are usually rectangular and can be made entirely of glass or can have opaque material on the sides with a clear top that lets sunlight through. Sometimes such boxes are called *Dutch lights*, based on the style popularized in Holland with windows that can be opened and closed for ventilation. A *hot bed*, by contrast, is constructed in such a way as to generate heat at the base of the plants. Plants grown in a hot bed grow very fast and strong from the warmth at the base. Individual plant protectors are called *cloches* and are made of either solid glass or plastic, often with some sort of opening in the top for venting.

For small rows or raised beds, a hoop structure of plastic sheeting is the most cost-effective. The hoops themselves can be made of heavy metal wire or pipe bent into a semicircular shape. Be forewarned if you use PVC—it is not UV-stable, and it will eventually splinter and break. Black ABS pipe works well, or you can get creative with salvaged items; for example, I have used flexible orange pipe left over from an underground wire

A cold frame with opaque sides and hinged tops.

installation, and I've also found that the wires used on rectangular campaign signs make great over-bed supports. They don't have to be semicircular, but a benefit of the rounded style is that, when covered with clear or semiopaque plastic sheeting, rain slides off the sides instead of puddling on the top surface and causing the covering to sag. Place the hoops (or other supports) over the bed, drape the plastic over the hoops, and secure the plastic at the bottom edges with something heavy or with stakes driven through the plastic and into the ground.

Long corrugated plastic panels, which can be rolled into a semicircular shape, are nice because they are light and can be stored flat when not in use. The types that have a UV coating can be expensive to buy new, but they last for years and years. I'm often able to salvage these.

For a more structural approach to a bed or planting area, you can build a cold frame from a variety of materials. I have stood old wooden windows on end and hinged them together at the corners; the structure can then be unfolded into a square or rectangle, depending on the sizes and orientation of the windows, and then folded relatively flat for off-season storage. With clear plastic or another window on top of the open structure, it's an individual greenhouse for a particular plant or bed. I created square cold frames out of salvaged shower doors placed on their sides to form a square, touching one another at the corners and supported by tall stakes on the inside and outside of each corner. I draped a clear plastic shower curtain over the top (and intermittently drained the puddle that collected). It wasn't pretty, but the peppers sure were!

When the days are short and the sun angle is very low, try to take advantage of as much light as you can with clear-sided cold frames. If you're getting plenty of light from above, you can do solid sides with a glass top. Plywood sides will swell and separate when they absorb moisture from the ground, but if you lay out a perimeter of bricks or rot-resistant wood (such as cedar or redwood) or synthetic materials, it will make a nice base for a plywood box. For over-plant protection, you will want at least an 8- to 10-inch-high foundation, preferably higher.

Solid side walls will help insulate the box. You can try backing the wood with Styrofoam panel insulation or even bubble wrap. Use your imagination! One year, I built a well-insulated bed enclosure from straw bales laid out in a square and shower doors across the top.

Triangular structures shed water effectively, but they limit the growing height at the lower corners if they are placed over a wide bed. Over a single row, they are great.

Cloches are among my favorite season extenders because they are small and usually, depending on the material, easy to store and manipulate. In the past, beautiful glass bell cloches were used extensively throughout Europe; you can still buy them, but they are incredibly heavy and expensive, and they become slick and difficult to grip when wet. Broken glass in the soil poses a serious risk to a gardener. The type of glass bell I admire has a small knob on the top; you carry it by using a special tool to grab the knob and pick up the bell. The biggest drawback is that glass cloches are solid and do not allow for ventilation.

Ventilation of cloches, or any season extenders, is very important. When covering your plants with anything, pay close attention to the sun, even if the outside air remains cold. The smaller the covered structure, the less inside air there is to heat, which means the temperature can quickly rise into triple digits. Without airflow, plants inside a clear structure in the sun can be quickly baked. Even with ventilation, too hot conditions can check the plants' growth, which is exactly the opposite of what you are trying to achieve by using season extenders.

I have seen plastic bell-shaped cloches with nifty vents in the top, but I need many for my garden and am unwilling to pay more than five dollars apiece. Years ago, I tried milk jugs with the bottoms cut off so I could set them over individual plants. Their lids could be removed for venting, and I could stack them easily and store them on a rope hanging from my rafters, but the square shape of their bottoms made them hard to sink down into heavy soil, and their light weight rendered them easily blown away. Also, the plastic degrades in UV light, so the jugs began to shatter and splinter after a couple of years.

Traditional cloches are pretty but can be expensive.

Many household items can be recycled and used as cloches.

The best things I've found to use as cloches are clear plastic half-gallon juice jugs. I saved up about a hundred of them, and I store them on ropes, like I do with the milk jugs. The caps are removable, and their round shape makes it easy to twist them securely into the soil. Best of all, they don't break down in sunlight, and they don't shatter if I drop them or toss them into a wheelbarrow.

You can use bed-size season extenders before there are any plants in the bed; a clear cover will heat the soil (and can be laid directly on the ground) and make it possible to plant seeds or transplants earlier than usual. Once you've planted the bed, suspend plastic sheeting above the new plants—but make sure it doesn't touch them, or else it may smother them or cause molding.

Storage of season-extension devices is often an issue. Because they are three-dimensional, they take up a lot of space when not in use if they do not fold or break down somehow. Plastic sheeting can be flattened and either folded or rolled up for storage. Support hoops can be stacked and stored in any out-of-the-way area. You can line them up on a peg or a hook on a wall or a fence; I hang mine from a rafter in my shed. Cold frames that don't disassemble or fold flat are a hassle if you have limited space. You can utilize them as storage areas, putting empty pots or plastic cloches in them during the warmer seasons, but it is easier if you can get them entirely out of the way when you are not using them. Bottom line: whatever style you decide on, make certain that your season-extension devices can be easily and safely dismantled or moved when you don't need them. If they are too difficult to contend with, you won't be motivated to use them, which defeats the purpose.

Once you grow your own food, you'll have a different appreciation for the variety of produce available year-round in an average grocery store. When you experience how long it takes to grow and ripen a bell pepper, or what sort of conditions cucumbers or melons prefer, you might start thinking about the many ways you can extend your growing season and how fun and rewarding it can be. This project is a simple design for a 2½-foot by 4-foot cold frame.

Build a Small Cold Frame

. . . .

Materials:
- [ ] **One sheet ½-inch plywood, 8 feet by 4 feet**
- [ ] **Twelve pieces rot-resistant 2x2 in the following lengths: two 12-inch, two 18-inch, four 24-inch, four 30-inch**
- [ ] **Clear corrugated roofing panel, 96 inches by 26 inches**
- [ ] **Four 2-inch utility hinges**
- [ ] **Rubber washer screws (and hex head bit for these, if needed)**
- [ ] **Thirty to forty 1¼-inch deck screws**
- [ ] **Eight 3-inch deck screws**
- [ ] **Drill with ¹⁄₁₆-inch bit for pilot holes and Phillips head for driving screws**

. . . .

Step 1: Build a wooden box. Cut a 12-by-48-inch rectangle across one 4-foot end of the plywood for piece A, and an 18-by-48-inch rectangle for piece B. Attach a 12-inch 2x2 upright to each short end of piece A with three screws. Next, attach an 18-inch 2x2 upright to each short end of piece B with four screws. Stand these pieces up (a helper or box to support them is useful) with the 2x2s to the inside. From the remainder of your sheet of plywood, cut a 30-inch square. Next, on one edge, mark off 18 inches from the left; on the opposite edge, mark off 12 inches from the left. Now take a straight edge and draw a diagonal line across the board between your two end marks. Cut the piece along the lines to create angled pieces C and D. Attach pieces C and D to the outermost exposed sides of the 2x2s, screwing them on with matching high (eight screws in all) and low sides (six in all). Congratulations! You just built most of the cold frame.

Step 2: Build the lights (i.e., the frame to hold the clear panel). You will create two rectangular frames, using two of the 30-inch 2x2s and two of the 24-inch 2x2s for each frame. Drill small pilot holes about ¾ of an inch in from each end through the four 30-inch 2x2s. These holes allow you to join the pieces of your rectangular frame in a series of butt joints. Drive one 3-inch screw through this pilot

hole and into the end of one of the 24-inch pieces, which should be butted up against the end of the longer piece (see illustration). Repeat this step until you have two rectangular frames. Measure these on the box as you cut and connect— you may have to adjust sizes because of the angle.

Step 3: Cut the corrugated roofing panel in half, widthwise, into two pieces measuring 48 inches by 26 inches. Lay the corrugated plastic panels over each wooden frame, putting the overhang at the bottom and outer side of each door. Attach the panel to the frame with the rubber washer screws, using three or four along each side, in the valleys of the plastic panel.

Step 4: Lay the lights on the top of the box, making sure that they are flush at the middle seam and that neither will obstruct the opening of the other. Attach the hinges onto the doors and the top of piece B (the back).

Note: You may want to have a tall stick at hand for times when you want to prop the lid open and reach inside with both hands.

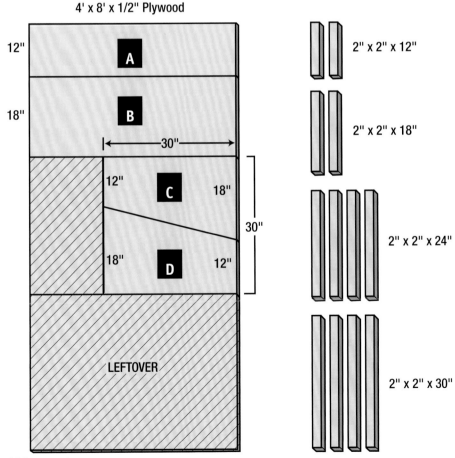

Lumber for cold-frame project.

Plant a Pollinator Garden

As more area is paved or built upon for human habitation, more insect habitat is destroyed—but we can all live together! If you plant some of the plants and shrubs discussed in this project for pollinators in your area, it could directly benefit your immediate food sources.

Why would you want to do this?
You realize that our food supply depends on natural pollinator species and that if they die out, we are next.

Why wouldn't you want to do this?
You have no place to plant plants, or you have an active cat (or neighborhood cats) who would use the garden as a buffet.

Difficulty level/skills needed:
Basic gardening skills.

Learn more about it:
The website of the Pollinator Partnership (www .pollinator.org) has information about pollinators nationwide as well as guides (and a smartphone app) that will help you find plants to attract the pollinators specific to your area.

A vibrantly colored aster attracts a visitor.

In the same way that every good reporter must learn about the "five Ws (and one H)" of journalism—who, what, when, where, why, and how—every good gardener must learn the "five Bs (and one M)" of pollination—bees, butterflies, birds, bats, bugs, and moths. There are many insects that assist in pollination simply by going from plant to plant and flower to flower (many flies help with this, whether we like it or not!), and our food supply depends on the success of pollination. So much of the food that humans eat relies on bee pollination, and there have been some alarming recent decreases in bee populations, especially that of native bumblebees.

By growing plants with flowers that are attractive to native pollinators, you are helping their populations, their survival rates, and, ultimately, our food supply. On top of that, the pollinators are fun and amazing to watch as they go about their day-to-day business, not knowing that so much depends on their biological imperative. The flowers are pretty, too. Here are some important points about starting a pollinator garden:

- All you need for a pollinator garden (besides the plants) is an area in which you can plant small flowering plants and/or shrubs. See Create a Garden Area (Project 1 of this section) for instructions on making your gardening space.
- Birdbaths are not just for birds—many insects use them to get water too. Make sure that your birdbath stays full and doesn't get icky with moss or algae.

- Leave seed heads in the winter for birds to graze on. Some people feel that this makes a garden look messy, but winter isn't optimal gardening time anyway, and the birds need it.
- The more diversity, the better. While big swaths of a single flower can be beautiful, insects and pollinators like to vary their diet just as we do.
- Consider purchasing binoculars if you don't already have them so you can enjoy watching the pollinators at work in your garden.

Goldenrod

154 Urban Farm Projects

Plants for a Pollinator Garden

Plant	Type	Pollinators Attracted
Abelia	Medium to large shrub	Bees, hummingbirds, bugs
Alyssum	Flowering ground cover	Butterflies, bugs
Aster	Small flowering plant	Bees, butterflies, bugs
Bachelor button	Small flowering plant	Bees, butterflies, bugs
Butterfly weed	Small shrub	Butterflies, hummingbirds, bugs
Cactus	Nocturnal flowering plant	Moths, bats
Chive	Clumping herb	Bees, butterflies, bugs
Clover	Low groundcover	Bees, butterflies
Columbine	Small flowering plant	Hummingbirds, bugs
Coneflower	Small shrub	Bees, birds, butterflies
Cosmos	Small flowering plant	Bees, butterflies, birds
Currant	Small shrub	Bees, birds
Dill (and family)	Herb	Bees, butterflies, birds
Evening primrose	Small flowering plant	Hummingbirds, bugs
Four o'clock	Small flowering plant	Bees, hummingbirds
Foxglove	Small flowering plant	Hummingbirds, bugs
Globe thistle	Small plant	Bees, butterflies, birds
Goldenrod	Medium shrub	Bees, bugs
Hyssop	Small shrub	Bees, butterflies
Lavender	Small shrub	Bees, butterflies, bugs
Lupine	Small flowering plant	Bees, hummingbirds, butterflies
Marigold	Small flowering plant	Bees, butterflies
Monarda	Small shrub	Bees, hummingbirds, butterflies
Moonflower vine	Annual vine	Moths, bats
Morning glory vine	Annual vine	Hummingbirds, bees
Passionflower	Flowering vine	Bees, hummingbirds, butterflies
Penstemon	Small shrub	Hummingbirds
Sage	Small to medium perennial or annual shrub	Bees, bugs, hummingbirds
Sausage tree	Tree	Bats
Sunflowers	Tall flowering plant	Bees, bugs, birds
Trumpet vine	Flowering vine	Bees, hummingbirds, bugs
Viburnum	Medium shrub	Birds, butterflies
Weigela	Medium shrub	Bees, hummingbirds, bugs
Wild lilac	Medium shrub	Bees, butterflies
Yarrow	Small flowering plant (annual) to medium shrub (perennial)	Bees, butterflies, bugs

Edibles on Your Deck

There are few activities that can be as rewarding as harvesting food that you've grown from seed in your own garden area. It's even more satisfying when you can grow your own food in a home without an actual garden. Start small, with simple expectations, and just give it a try.

Why would you want to do this?
It's a small-scale project that's easy and fun and yields tasty results.

Why wouldn't you want to do this?
You don't like to eat homegrown food, you don't cook, you have no sun exposure for growing plants, or you travel too much to care for plants.

Cost comparison:
You can lower your produce costs if you replace store-bought ingredients in your favorite dishes with ingredients you've grown yourself.

Skills needed:
Basic gardening skills.

A corner of your deck can be useful and beautiful.

Steamed or roasted vegetables? Green beans, carrots, onions, beets. Fresh snacking vegetables? Carrots, kohlrabi, beans, peas, radishes. Salads? Lettuce, arugula, spinach, beets, scallions, cucumbers, carrots, tomatoes, summer squash, corn, peas, bean sprouts, edamame, cauliflower, broccoli, quinoa.

Materials:
- ☐ **Deck or balcony with sun exposure**
- ☐ **Pots/planters**
- ☐ **Soil**
- ☐ **Seeds**

Step 1: Come up with a list of specific recipes/dishes.

Example: I love my mother's summer cucumber salad recipe, my husband makes wonderful sautéed green beans, and I also like steamed carrots. All three of these dishes contain either onion or garlic.

Step 2: Look at your list and decide which ingredients can be easily grown in your space to yield the amounts you need.

Example: From my list above, I can grow cucumbers, red onions, dill, green beans, garlic, and carrots.

Step 3: Figure out which of your plants, if any, can be grown together in a single planter and determine how many planters you'll need.

Example: Beans and carrots are friends when planted together (see the Companion Planting sidebar on the next page), as are cucumbers and dill. The onion family can be either very helpful or very problematic as companions, so I'll plan a separate container for my onions and another for my garlic. I'll need four planters.

Step 4: Obtain/prepare your planters. Decide which vegetables need the most room for roots and which ones need the most above-ground space. How you can provide this most efficiently in the available space?

Example: I'll put my beans and carrots in a large planter because they will need the root room, and I want lots of them. The green beans can grow up strings, a trellis, or a net on a wall, leaving room at the base for the carrots.

In books about gardening for or with kids, authors often point out that children will more readily eat vegetables they've helped produce. It's suggested for parents to make sure that children do eat what they grow; otherwise, they don't get the full experience of producing their own food. I've also seen suggestions such as "grow a pizza garden," a project in which kids plant what they like on their pizza (e.g., tomatoes, peppers, basil, spinach) and then, at harvest time, the kids and parents build the pizza together. The pizza idea is a little challenging for me because my kids couldn't grow the extra mozzarella, pepperoni, or black olives that top every pizza in our house, but I like the idea of choosing a favorite dish and planting a garden that can provide you with many of the ingredients for that dish. It's a perfect project for both neophyte gardeners and urban dwellers who are able to grow only in containers.

What is your favorite dish, and what ingredients for it can you can grow? Ratatouille? Eggplant, squash, tomatoes. Potato salad? Potatoes, celery, green onions.

Companion Planting

The *Oxford Dictionary* defines companion planting as "close planting of different plants that enhance each other's growth or protect each other from pests." Many plants emit chemical messages that can either agree or disagree with those from other plants; thus some plants coexist with each other well, while others do not. Some common pairings to consider (or avoid):

- Parsley grown with asparagus improves the strength of both.
- Basil and tomatoes are an age-old partnership in the garden as well as on the plate.
- Carrots and/or radishes paired with pole beans fit space well and improve one another's vigor, but beets at the base of pole beans can be a detrimental to both.
- Bush beans paired with strawberries or cucumbers boost the growth of each.
- A traditional Native American companion setup, also known as the "Three Sisters of Life," is corn, pole beans, and winter squash vines. The beans use the cornstalks for support while contributing needed nitrogen to the corn. While the squash happily crawls around the base of the others, benefitting from the partial shade and wind reduction, its prickly stems and leaves seem to deter corn-loving pests such as raccoons and mice.
- Even though carrots and dill have similar soft, feathered greenery, when planted in combination, each reduces the strength and size of the other.
- Dill interplanted with onions is a pretty and mutually beneficial combination.
- Onions are notorious for impeding the growth and vigor of legumes such as peas and beans.

Cucumbers also like root room, but they aren't as aggressive as beans, so they can go in a smaller pot, also with a trellis to climb on. The dill is light and feathery and can grow thickly at the cukes' base.

You can do a long gutter or a deep, long window box with a line of garlic right down the middle (start in the fall for a mid-summer harvest). A big square or rectangular pot will work for the onions. They don't mind being a bit crowded, so just figure out how many full-size onions will fit in your pot, allowing for a little space between each one.

Step 5: Use the vegetables! Many plants stop yielding produce if they remain unharvested. Pick the produce as it ripens (even when young—baby vegetables are gourmet delicacies!) and enjoy your homegrown flavors.

There you go: four planters, the makings for at least three great vegetable side dishes, and very little floor space consumed. As you harvest, fill in the empty space with more carrots for an ongoing supply. If you feel inspired, do more containers: lettuce, arugula, and spinach for salads; and herbs in a window box hanging from a railing or on the front porch.

Creative use of space on a deck or patio area will maximize how much you can grow.

Make Your Own Planters

Even if you don't have a yard or ground space, you have no excuse not to try gardening and growing your own food. Many apartment and condominium dwellers have gardens of their own, and I've been to restaurants that grow their own produce on the roofs of their buildings. With some creativity, there are many possibilities for people whose homes have no outdoor space.

Why would you want to do this?
You want to garden or try to grow your own food but have no yard or no place to dig in the ground.

Why wouldn't you want to do this?
You have nowhere to put large pots or planters outdoors.

Is there an easier way?
You can buy a large pot (or pots) from a nursery or hardware store and fill it with soil; this will be much easier but will also be much more expensive and possibly not as durable.

Cost comparison:
Large store-bought planters of the size discussed in this project can be very expensive. When you make your own planters, the biggest expense is that of the potting soil.

Skills needed:
Basic construction skills and the ability to do a bit of heavy lifting.

Learn more about it:
McGee & Stuckey's Bountiful Container (Workman, 2002) by Rose Marie Nichols McGee and Maggie Stuckey.

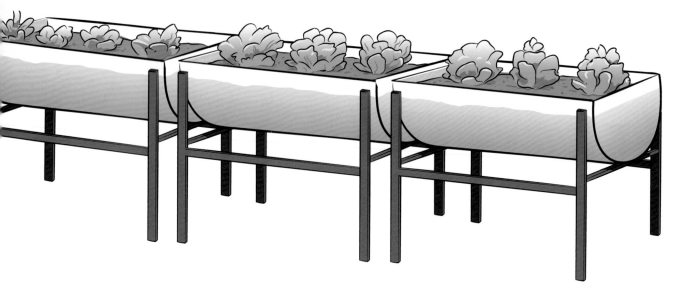

The list of containers that can be used as planters and materials that can be adapted into use as planters is a long one. In Project 6 (Vertical Growing) of this section, I talk about using gutters for planters; I have also drilled drainage holes in extra recycling bins to convert them into planters. Many people recycle tires by utilizing them to grow potatoes: fill the center of an old tire with soil or mulch, plant your potatoes, and then stack on another tire (and add more soil or mulch) as the plant grows. I've used metal 55-gallon drums, both whole and cut to half height, and I've turned a whole straw bale into a planter, which is a lot of fun—but a lot of mess for an apartment dweller to be left with at season's end. In England, many urban dwellers use big bags of soil intact, planting through holes or slits cut directly in the plastic (there are even special devices to keep the holes open and help with watering). I began pondering the wide variety of options when a restaurateur for whom I grew produce told me about a friend who grew carrots in big plastic kiddie pools.

...alf either vertically or horizontally, depending on how much growing area and depth you want in your planters.

For our front porch, I purchased nice pots, wanting my citrus trees showcased with a Mediterranean feel. Large ceramic pots are usually the best looking but also the most problematic. In addition to being heavy and expensive, terra-cotta pots can easily split or break. If they are unglazed, the pots breathe, which means that keeping the plants from drying out is a challenge. Plastic pots are cheaper and lighter than terra-cotta; however, they not only lack quality but also are not UV-stable and can fade and splinter over time. There are rigid Styrofoam-like pots made to resemble stoneware or terra-cotta—I tried these on our upstairs deck—but they dent and chip when bumped and don't look so attractive after a couple of years.

You can use wine or whiskey half-barrels as planters, which give a size and volume comparable to my project, but the wooden staves will eventually rot out and loosen, and the barrels are hard to find cheaply anymore.

This project uses a big plastic barrel and gives some leeway as to the orientation. When you cut the barrel in half horizontally, you'll have two circular planters, each with 3 square feet that can be thickly planted because of the 17-inch depth. Cutting the barrel vertically, I created two shallower planters with more area (almost 6 square feet) but less root room (my carrots love them) and a semicircular profile that maximizes drainage.

Materials:
- [ ] **Food-grade plastic 55-gallon barrel**
- [ ] **Indelible marker**
- [ ] **Saw (I recommend a reciprocating saw; try to borrow or rent one if you don't have one)**
- [ ] **Drill with ¼-inch or ⅜-inch bit**
- [ ] **Piece of mesh, screen, or landscape cloth**
- [ ] **Four bags of potting soil**
- [ ] **Eight bricks**

Step 1: Decide how you want to cut the barrel: either around the center, at its "waist," or down the middle, from top to bottom.

Step 2: Using an indelible marker, draw a line that you will follow when cutting. It doesn't have to be straight—you can get creative with curves if you want.

Step 3: Cut the barrel along the line with your saw (you'll see why I recommend a reciprocating saw).

Step 4: Drill holes in the bottoms of both halves of your barrel. If you've cut your barrel from top to bottom,

Half of a barrel that was cut across its "waist."

Drill drainage holes all over the bottoms of the barrel halves.

leave the bungs (stoppers) in place and drill a single line of holes about every 2 inches at the deepest point of each half. For the bottom half of the horizontally cut barrel, drill a smattering of holes all over the bottom, especially in any areas where water could collect. For the top half, unscrew and remove the bungs and then drill holes all over the rest of that end, again focusing on areas where draining water would collect.

Step 5: Flip both halves so that the open sides face up. Move the empty planters to their intended locations.

Step 6: Place four bricks in a square or rectangle under each empty planter, making sure that the planter is stable. The bricks secure the planters from rocking as well as raise them to allow water to drain out.

can act as "risers" if you don't have bricks.

Step 7: With a horizontally cut barrel, place a small piece of mesh, screen, or landscape cloth over the big holes where the bungs were to keep your soil from spilling out.

Step 8: Fill the planters with potting soil, which contains ingredients to keep it lighter and prevent soil compaction. For my carrots, I used a very sandy mix, and the planters are impossible to move now.

Step 9: Plant and water.

Additional Advice

- If you are concerned about bending over to plant or harvest, build a sturdy stand for your planters so that they are higher and easier to work with. I welded metal supports for mine.
- You can paint the planters if you'd like to dress them up a bit; there are brands of paint made specifically for plastic, but I've found that some still flake off in time.
- You can disguise the planters by hiding the plastic with foliage: cluster other plants in smaller pots around the planters, or plant trailing plants at the planters' edges to drape over the sides.
- Make your planters look less like barrels or hide the containers entirely by wrapping them in an attractive material, such as a piece of bamboo privacy screen, a strip of salvaged linoleum, or aluminum flashing. You can adjust the appearance of the planter, but you can't beat the functionality!

Get crafty with the outside of your barrel to give it aesthetic appeal and match it with the rest of your garden decor.

Vertical Growing

If you consider how many more humans can live in a given area of square footage when residences are built skyward, doesn't it make sense to apply the same concept to growing plants?

Why would you want to do this?
Why not make the most of the space you have to grow food in?

Why wouldn't you want to do this?
You don't want to grow plants, or you don't have a place to hang heavy planters.

Is there an easier way?
There are some versions of "stackable" planters available, but I haven't seen any others that hang like this.

Skills needed:
Basic construction skills.

Learn more about it:
I haven't found many resources for similar projects; most books in the "vertical gardening" category deal with trellising and supports to attach trailing plant growth to. If you look up "vertical farming," you'll find information that's closer to what I'm thinking about, but on a much larger scale.

Raised Beds

As demonstrated by most of the projects in this section, a garden doesn't have to go in the ground. Technically, plants need to be in soil (excluding hydroponic growing), but where and how that soil is located is not absolute. By building a raised bed or two, you can create a garden wherever you want one and increase your success by controlling the soil quality and condition.

Why would you want to do this?

Raised beds can extend your growing season because the raised sides collect early spring sun and warm up before the ground does, giving most plants and seeds a jump start. Gardening in raised beds can be easier on backs and knees.

Why wouldn't you want to do this?

You don't want to build a structure, you don't have room or sun exposure for a several-square-foot container filled with soil, or you have nowhere onto which water can drain.

Is there an easier way?

Have the hardware store cut your lumber to the desired lengths for you, and ask someone (like that good-looking neighbor!) to help you attach the pieces and move the bed into place.

Cost comparison:

Following these instructions to make a raised bed costs far less than using a raised-bed kit or buying a premade bed.

Skills needed:

Basic construction knowledge, such as how to drill a hole and use an electric drill to install screws (nails can loosen and pop out as the wood shrinks and swells over time).

Learn more about it:

Square Foot Gardening (Rodale Books, 1981) by Mel Bartholomew is one of the classics of small-scale and raised-bed gardening, and it will give you tons of great tips and ideas on the many plants you can grow and how to grow them in a relatively small space. It is a bible for many raised-bed gardeners.

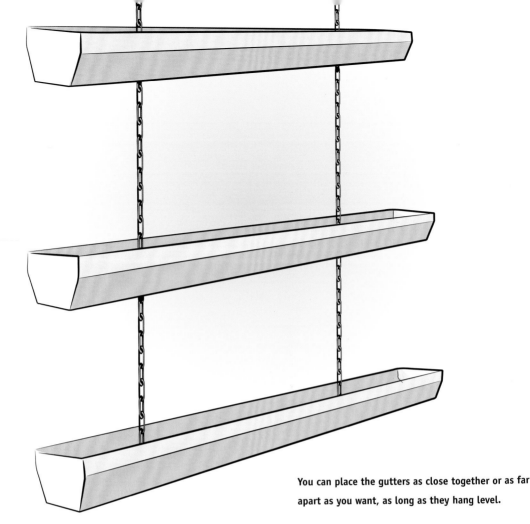

You can place the gutters as close together or as far apart as you want, as long as they hang level.

with a washer on the outside bottom. Screw the center piece of the turnbuckle to the protruding threads.

Step 4: Put the two eyehooks through the gutter piece with two holes, again with the eyes inside and the washers bolted to the outside between the gutter and the nut.

Step 5: Using the chain or cord, connect the eyehooks inside the gutters to an even length of chain for each side of the gutter. Make sure that the height between each gutter is sufficient for what you are going to plant and is enough so that the gutters don't shade one another; about 12 to 15 inches should be fine. Connect the hook beneath the gutter length to a piece of chain, and use an S hook to connect it to the lower eyehook. Do this again for the final gutter piece, which should be the undrilled piece with only eyehooks in it.

Step 6: Hang the whole apparatus from the ceiling or roof. Make sure that all of the gutters are level.

Step 7: Fill each of the gutters with soil, tamping down gently.

Step 8: Plant your seeds or starts.

Additional Advice

- This planter is a good size for shallow-rooted plants and is easy to use for succession planting, picking from one level while the other two grow.
- Try different varieties of lettuce or other greens that grow quickly and stay relatively small. You can harvest most greens when they are small—they are very sweet and tender delicacies—or you can harvest every other head at that time, enjoying some baby greens while making room for the others to continue growing.
- Water the top level of the planter heavily for extra outflow to the lower tiers (this will work if you have used proper potting soil and drilled holes as per the instructions) or just water each level individually. If you didn't drill a line of holes all the way across the lowest gutter, be careful that the lowest level doesn't get too wet because it has no drainage.
- Look around your house for other vessels that can hold soil to make variations on this planter and expand your options for small-scale vertical gardening.

Vertical Growing **169**

library

6/11/2018 3:58 PM
Tx ID: 3M 2UA2281N1Z0000008859

Customer Name:
Calacci, Christina L.
Customer ID: ************5537

Type: Payment
Amount: $18.40 USD

************5942

Visa

Status: Approved
0

Approved Amount: $18.40 USD
Total: $18.40 USD

Signature

Mode: Issu d

AID: A0000000031010
TVR: 8000008000
IAD: 0601A03A00000
TSI: 6800
ARC: 00

Thank you for visiting us!
ballibrary.org

library
barrington area

6/11/2018 3:58 PM
Tx ID: 3M-2UA2281NLZ0000008859

Customer Name:
Caiacci, Christina L.
Customer ID: **********5537

Type: Payment
Amount: $18.40 USD

************5942

Visa

Status: Approved
0

Approved Amount: $18.40 USD
Total: $18.40 USD

Signature

Mode: Issuer

AID: A0000000031010
TVR: 8000008000
IAD: 06010A03A00000
TSI: 6800
ARC: 00

Thank you for visiting us!
      balibrary.org

If you think about it, we stack a lot of things: books, dishes, apartments, boxes, cars, clothes. Stacking is one of the more effective ways of using space efficiently. We don't often think of stacking plants, but if humans can live in stacked houses, why can't plants? We live in three dimensions, but conventional gardening seems to deal in only two dimensions, stuck in the ground.

I'm not sure if the macramé craze of the 1970s is what spawned hanging planters, but that ecru-colored weaving with funky wooden beads seemed made to showcase trailing plants (other than belts, what else could it be used for?). Hanging planters turned unoccupied air space into usable garden areas.

Hanging planters have become ubiquitous, and they provide a fun and easy way to decorate otherwise not-plant-friendly areas. But what about using them for growing food? A few years ago, a friend grew cherry tomatoes in a hanging pot on her deck, and she was surprised by how many people remarked on "such a clever idea!" Since then, the upside-down tomato planter, which uses gravity for watering, debuted, cementing the idea of growing a food crop from a hanging position.

Installing thin light fixtures above the tiers of hanging planters improves the plants' growth. You can turn any bookshelf into a series of plant shelves by adding inexpensive shop-light fixtures with grow bulbs. When you use shelf space for gardening, you multiply your growing area without changing your existing square footage.

As an example, I have a small 8-foot-by-10-foot greenhouse with 35 square feet of table space and a 12-square-foot inground bed. Over the inground bed, I built a fold-down counter (9 square feet) and four shelves (6 square feet each), and I put a 10-foot-long 6-inch shelf above my tables. By stacking growing areas in the open vertical space, I netted an additional 38 square feet, effectively doubling the usable space but retaining the same footprint.

But here's even more! We had old recycle bins that hooked on an upright aluminum wheeled cart for transport to the curb. I drilled holes in the bottoms of the bins, filled them with a light soil mix, and planted them with early carrots in the greenhouse. In about 2 square feet of floor space, the cart served as a stand for two carrot planters and a way to transport them outside when the weather got nice.

My favorite vertical planter is described in the following project. I've made several different versions of it, all of which I have used in my greenhouse.

This planter is intended to hang in an outside area with good sun exposure, although it is possible to hang it indoors. The most important factor is that whatever you hang it from must be securely mounted into the ceiling or roof and able to support the planter's weight—with its multiple levels, this planter is very heavy when full of soil, even more so when wet. The plants start out as seeds, but they also add weight as they grow larger.

Materials:

- ☐ Three plastic gutters, as deep as possible, cut into 2- or 3-foot lengths (see if you can salvage these)
- ☐ Six gutter ends that fit on the gutters
- ☐ Four $^3/_{16}$-inch hook-and-eye turnbuckles
- ☐ Two $^3/_{16}$-inch-by-$^5/_8$-inch eyehooks with washers and nuts
- ☐ Four $^3/_{16}$-inch strong washers for turnbuckle diameter
- ☐ Six hooks
- ☐ Drill, with $^3/_{16}$-inch or $^1/_4$-inch bit
- ☐ Strong utility chain or weight-bearing cord
- ☐ Potting soil
- ☐ Seeds

Step 1: Secure the gutter ends to the individual gutter pieces. You don't have to caulk or seal them if they fit snugly enough; most gutter caps are designed to hold water inside. If you're not sure, fill the gutter up with water to see if it leaks with the caps on.

Step 2: Turn two of the three gutter pieces upside down and drill a line of holes down the length of the midline, about 1 inch apart. Be sure that there are holes at the 8-inch and 16-inch marks if using 2-foot gutter segments or at the 12-inch and 24-inch marks if using 3-foot pieces. In the bottom of the third piece, drill only two holes, either at the 8- and 16-inch marks (for a 2-foot segment) or the 12- and 24-inch marks (for a 3-foot segment), centered on the width. (Or, if you plan to hang the planter in an area that can get wet, drill all three gutters with a full line of holes.)

Step 3: Attach the turnbuckles through the 8-inch/16-inch or 12-inch/24-inch holes in the two multiple-drilled pieces. Put the eye side through the inside of the gutter

No matter where you live, growing in raised beds is a great way to garden. If you have a yard, you can place the bed almost anywhere that has good exposure and skip the heavy-duty work of digging, which earthworms and roots will do naturally over time once the bed is in place. If you're a condominium dweller, you can create growing areas where there is only hardscape, or you can even bring your planting area onto a sunny deck or balcony. If you live in an apartment, check with the building manager about constructing beds on the roof—with proper drainage, the roof is an ideal, yet often under-utilized, space for gardening.

Soil conditions inside a raised bed can be easily kept optimal for plants. By relegating compaction from footsteps or wheelbarrows to external areas, you can make sure the contained soil stays aerated for best root growth and health as well as good drainage. The beds can be maintained by simply topping them off with compost or mulch, and everything stays tidier because all of the dirt is contained.

Before obtaining materials, decide where you want the beds to be and how large an area you want to grow in. Although their sides can be as high as you'd like, raised beds should not be so wide that you can't reach the middle—for planting, harvesting, and maintenance—from either side. Because this is a bottomless box, it should be positioned only on a surface that will tolerate direct soil contact and absorb or channel water completely. Do not situate it on a wooden deck, porch, or balcony.

...raised beds can turn a paver patio into

Deciding on lumber is often complicated because of the many types that exist. In our area, cedar, although expensive, is the most common choice. In California, redwood is by far the most prevalent. Both of these woods have a natural resistance to rot and decay, which is imperative because they will be filled with dirt and water for their entire useful lives. If you use pressure-treated wood, make sure that the preservatives used do not contain arsenic if you are growing edibles. Otherwise, line the beds with plastic to prevent the toxins from leaching into your growing medium. Another option is a composite "lumber" product made from a mix of sawdust and recycled plastic. This is more expensive if bought new, but it will last indefinitely and can be curved into interesting shapes for aesthetic effect.

Materials:

- ☐ 2x6 or 2x8 lumber of desired lengths, allowing for overlap on ends
- ☐ Electric drill
- ☐ Drill bit
- ☐ Phillips head deck screws
- ☐ Phillips head screwdriver bit
- ☐ Plastic for lining bed, if desired (you can use salvaged plastic bags or sheets)
- ☐ Cardboard, newspaper, or landscape cloth large enough to cover the bottom of the bed
- ☐ Soil

Step 1: Cut the lumber into the desired lengths for your bed walls, accounting for overlap at the ends.

Step 2: Attach upright supports to sides longer than 6 feet to prevent the sides from bowing when full of soil. Supports can be as simple as a 2x2 or 2x4 screwed into the middle of each long side (drill a pilot hole for each screw to prevent the wood from splitting during assembly). If the bed will be placed on a soft surface, such as dirt, the supports can extend below the bottom of the side pieces to be anchored into the ground.

Step 3: Drill pilot holes at the edges of the side boards where the screws will go. With a helper, screw the side walls to the ends of the end wall pieces.

Raised beds are a good option if you want to turn designated areas of your deck into gardening spaces.

Step 4: With a helper, carry the bed to its intended location. By having a "free-floating" bed, you can move it around and adjust it until you are happy with it. It can be left as is, and the weight of the wood combined with the filling will keep it in situ.

Step 5 (optional): If you are putting the raised bed in a yard, attach metal pipes or wooden stakes for additional support and to prevent shifting of the finished bed. To use metal pipes (I salvage them), sink them inside the frame and secure them to the walls with pipe clamps. (These pipes can also be used to support hoops or trellising over the bed.) If you prefer to use wooden stakes, make sure that they are of a rot-resistant variety, and screw the bed walls directly to them from the outside.

Step 6: Line the bottom of the bed with cardboard, newspaper, or landscape cloth. Cardboard and newspaper will block grass and prevent weed seeds from germinating; it will eventually rot, allowing in-bed plants more root room. Landscape cloth keeps out the really pernicious weeds, allows drainage, and holds in the soil, but it can impede the roots of larger plants in the bed.

Step 7: Fill the bed. You can start with manure or unfinished compost, which will eventually rot, providing roots with a nutrition boost, or you can fill the bed with planting soil or finished compost.

With regular mulching, you can keep the planting beds filled and will see a happy, thriving population of earthworms and other beneficial organisms that will perform the tilling for you so you can have all the fun of planting and growing.

A bed can be placed on top of a grassy backyard area with less-than-ideal soil conditions.

Growing in Tiers

If you have the space for a small raised bed or some large planters, you can grow different types of plants, including taller plants with deep roots, without giving up any extra floor space.

Why would you want to do this?
You like growing, but you have limited space and want to make the most of it without taking away square footage.

Why wouldn't you want to do this?
You are content with your garden space and feel that you have room to grow everything you'd like.

Skills needed:
Basic gardening skills plus some creativity in planning out what to plant in each tier for aesthetic appeal.

Further refinements:
Last summer, while preparing to go away for a week's vacation, I needed to quickly maximize my pre-existing automatic watering setup on the patio. I grabbed some of my small pots and stuck them into bigger pots with drippers in them. I extended the watering time so that the small pots would get more water, which would then flow directly from their drainage holes into the larger pots. Using the pots in tiers was a quick solution.

I also recently saw a wide staircase made from landscape timbers. It was maybe 15 to 20 feet across, and at least eight steps high, and someone had used the structure for tiered planting. The gardener had planted taller grasses and small shrubs in a pattern, similar to how one would lay bricks so that no mortared seam was directly above a lower one. It forced the person climbing up the stairs to walk in a zigzag so as not to step on any of the plants. It was an interesting and beautiful use of a necessary but repetitive landscape element.

Where hanging is not a good option, stacking up from a solid base is an alternative. It is possible to build a vertical planter by piling progressively smaller planters on top of each other. With bottomless containers, you can create a tall, deep space for a larger plant or tree, retaining the lower, shallower areas for the smaller and less aggressive plants that might otherwise get overrun.

Consider what you like to eat that you could grow in small planters, and how little floor space a stack of those planters could fit in. Planting in tiers is little more than combining vertical gardening (Project 6 in this section) with a raised bed (Project 7 in this section) with one key difference: instead of hanging the planters and medium in their own individual containers, you're stacking planters of successively smaller diameter with the largest, heaviest one serving as a foundation. This is a boon for growing because you gain root depth in the higher beds or planters with the stability of a wide, strong base.

When you plant tiered beds/planters, you have different options and dimensions than you would with a single-depth planter. You can put shallower-rooting plants in the lowest level and larger/deeper-rooting plants in the middle and upper tiers. Another option is to put trailers and creepers (such as strawberries) in the lowest tier and more upright or taller plants in the middle and uppermost tiers so that they don't shade the lower ones. Yet another plan is to put taller plants in the lowest level specifically to contribute shade to the upper levels, and then grow shade-tolerant plants, such as spinach and lettuce, in the upper tiers.

Tiers with Raised Beds

Materials:
☐ **see Section 4, Project 7 (Raised Beds)**

Step 1: Follow the steps in the Raised Beds project to build a raised bed of the desired size and dimensions to serve as the base. Fill it with soil.

Step 2: Decide how much space in the middle or back of the bed you'd like the next tier to occupy. Build another bed, as before, with these reduced dimensions.

Step 3: Tamp down the soil in the first bed, at least where the next tier will sit, so that the upper beds have a solid foundation that won't shift much as the soil settles.

Step 4: Place the second tier on top of the first and fill it with soil.

A tiered kitchen garden supplies herbs and leafy greens.

Step 5: If you'd like a third and fourth tier, repeat Steps 2–4 with smaller dimensions as desired. Four is the maximum number of tiers recommended for this project.

Step 6: Plant the levels as desired.

Tiers with Planters

- -

Materials:
☐ Three or four pots, either with drainage holes or with the bottoms cut off, each a few inches smaller in diameter than the previous one

- -

Step 1: Fill the largest pot with soil.

Step 2: Set the next largest pot on top of the soil in the largest pot, and fill the second pot with soil.

Step 3: Repeat Steps 1 and 2 for the third (and fourth, if applicable) pot.

A whimsical take on using planters in tiers.

Build a Trellis

Similar to how a large number of people can live in a small footprint in a high-rise building, a number of plants can be grown in a small bed if they are encouraged to grow upward. Upward growth can be accomplished with structures of wood, metal, or even string; for many plants, climbing is a natural growth pattern and is healthier for them than growing across the ground.

Why would you want to do this?

You can gain growing space, thus allowing you to grow more, and you can create or emphasize microclimates with these "living walls." As a bonus, the flowers on flowering plants will be more visible and may attract hummingbirds.

Why wouldn't you want to do this?

You don't have any plants that want to grow tall, you don't want tall structures in your growing area, or you have constant high winds or risk of lightning in the trellis area.

Is there an easier way?

You can grow climbing plants on existing structures: deck and stair rails, downspouts, conduit and pipes on exterior walls, laths over crawlspace areas. You also can purchase prefab structures—they are widely available in many different styles, but they often are not very durable. Making your own trellis will give you a strong structure for less money and allow you to be creative, perhaps even matching patterns that already exist in your home or garden.

Cost comparison:

With a homemade version, you can salvage many of the materials, thus getting them for free. Even if you purchase the basic materials, it will be much less expensive than buying a ready-made model.

Skills needed:

Basic carpentry skills.

When a plant has a climbing habit, it is unhealthy for it to sprawl on the ground, touching the soil. Soil contact for many plants can encourage fungus, mold, or mildew, which will inevitably cause rot if not remedied. By minimizing the plant's footprint, it also minimizes paths that destructive insects can use to get to the plant, exposes more of the leaves to sun and air circulation, and gives birds and other pollinators better access to the flowers. Many plants will act as natural screens for providing summer shade or windbreaks, and fruits on trellised plants are better shaped and more evenly ripe than those grown on the ground. Some fruits may ripen earlier due to the increased exposure to light, warmth, and airflow.

Materials:

- ☐ Cedar (or other rot-resistant wood) 2x2, 6 or 8 feet long (use the longer length for softer soil)
- ☐ Cedar (or other rot-resistant wood) 1x2s in the following lengths: one 2-foot piece, one 3-foot piece, one 4-foot piece, and two 5-foot pieces
- ☐ One 5-inch galvanized or stainless steel bolt, with nut and two washers
- ☐ Nine 1½-inch wood screws (or nine 2¼-inch thin bolts, each with two washers and nut)
- ☐ Drill
- ☐ Screwdriver
- ☐ Pliers

Plants with an Upward Growth Habit

EDIBLES	EASY CLIMBING ORNAMENTALS
Beans	Cardinal climber
Cucumbers	Climbing hydrangea
Espalier tree (fruit that grows on a trellis)	Honeysuckle
Grapes	Morning glory
Kiwi vines	Sweet pea
Melons	Trumpet vine
Peas	
Raspberries	
Tomatoes	

Ivy is a natural climber and a favorite for walls and trellises.

Trellises along the fence provide a living backdrop for this garden.

Note: Before you begin, it is helpful if you cut the bottom of the 2x2 at an angle, or, better yet, angle it on all four sides because it will be pounded into the ground. Also, if you'd like the "arms" of the trellis (the two 5-foot pieces of 1x2) to be closer together, cut the crosspieces shorter than 2, 3, and 4 feet; to make them spread wider, cut the crosspieces longer.

Step 1: Drill a hole of the same diameter as the 5-inch bolt all the way through the 2x2, about 3 feet up from the bottom. Drill another hole of the same size in one end of both 5-foot pieces of 1x2, about 1½ feet up from the bottom.

Step 2: Place both 5-foot pieces of 1x2 on top of the 2x2, lining up the holes in all three. Put the 5-inch bolt through all three pieces, making sure there is a washer on both sides of the wood. Attach the nut to the bolt, but don't fully tighten it yet.

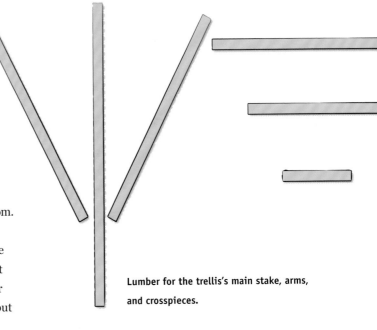

Lumber for the trellis's main stake, arms, and crosspieces.

Other Materials

You can use other types of materials for trellising, depending on your imagination and what sort of plants you're growing. For example:

- Thin pipes or tubing, such as galvanized (will rust, which can be pretty) or copper (will oxidize)
- Strong fence panels, such as livestock fencing (comes in panels with gaps of various widths) or regular fencing (salvaged or found at hardware stores and cut to size)
- Concrete reinforcing wire, available in panels at hardware stores or salvaged (has 4-inch or 6-inch openings)
- Old-style bed or cot frames, which are metal rectangles with holes along the sides that can be used to attach support strings (they can often be salvaged)
- Bamboo pieces, tied together with twine or wire into a square or rectangular pattern

Step 3: Lay the wood down on the ground and splay the 1x2 pieces out at angles from the center post.

Step 4: Lay the three shorter pieces (crosspieces) of wood on top of the splayed 1x2s at even spacing, adjusting as you lay it out. Start with the 2-foot piece on the bottom, near the fork where all three pieces meet. The 3-foot piece will go in the middle, and the 4-foot piece will be the highest.

Step 5: Once you are happy with the placement of the crosspieces, drill either a pilot hole at each intersection and screw in a wood screw (this method is faster), or drill a hole all the way through the overlapping pieces and attach a bolt (this method is stronger).

Step 6: Tighten the main bolt at the base (and the smaller bolts, if using them).

Step 7: Stick your trellis into the soil just behind where you will grow a climbing plant. For added sturdiness, use a mallet to pound the center post in from the top.

When you are happy with the placement of your trellis, grow a climbing plant and watch it reach up and stretch along its arms.

Irrigation Systems

Although hand-watering is the best way to keep tabs on your garden and plants, it isn't always possible to water your plants as often as they need it. There are alternatives to hand-watering that don't require a plumber or expensive in-ground ditch-digging.

Why would you want to do this?
To save time, to make sure that you're meeting your plants' watering needs, or to water your plants at the times that are best for them but not ideal for you (such as early on weekend mornings in the summer).

Why wouldn't you want to do this?
You don't have that many plants, or you are in a groove with hand-watering and don't mind doing it.

Is there an easier way?
Getting advice makes the process easier. Before you get started, talk to the employees at your local hardware store to get their input on the best way to implement an irrigation system for your particular setup.

Skills needed:
Irrigation plumbing of this sort is quite easy and designed to be installed by the homeowner.

Learn more about it:
Find more info online: Irrigation Direct (www.irrigationdirect.com), click on "Expert Advice"; Castle International Resources (www.hydrosource.com), after clicking on "Library," look at the article titled "The Chapin Bucket Irrigation Kit"; and "Do It Yourself Bucket Drip Irrigation" (www.csupomona.edu/~jskoga/dripirrigation), an article by science librarian James S. Koga of California State Polytechnic University, Pomona.

A drip emitter on an overhead length of pipe.

When I first started my garden, I had no water access (that I could find, but that's another story). I was gung ho on rain barrels and proudly announced that I would be watering my half-acre garden with the water that the rain barrels (downhill from the garden, at the house) collected. Carrying watering cans 300 feet each way got old pretty fast. Attempting to pull a full rain barrel uphill in a cart was sheer idiocy that I recognized as soon as I, yes, tried it. That year's garden was a little thirsty. The water from the six rain barrels I installed at the garden shed lasted a few weeks during the summer, but the watering-can method was tedious. A very long hose from the barrels was great—as long as the end of the hose stayed lower than the barrels. If I raised it up above my knees, the flow of water stopped. So much for my sustainable-water harvesting! I broke down and had a dedicated valve tapped off of our house's main line. I still dream of digging a well, but that cost versus the current cost of municipal water doesn't balance out.

Although I have many glowing childhood memories of watching my father hand-dig deep trenches for his in-ground sprinkler system, I have always been too fond of my back to emulate him. Instead, I use an above-ground method that I shut off and drain each winter; it isn't very pretty, but I've never regretted it. I've made a number of changes and adjustments as needed over the years, and it's an easy thing to do.

Before explaining the steps of this project, here are brief explanations of some of the materials that you'll find in the list that follows. There are a few basics you'll need for any system.

- A female end to connect the entire system to the faucet.
- A screen washer to put into the female connector at the faucet.
- A backflow preventer, which generally is legally required. Some of the faucet-mounted varieties squirt, so watch out!

- A timer. Timers can range from simple dials to elaborate programmable models, complete with rain gauges and seasonal adjustments. Get the simplest one you think you'll need. If you're forgetful, get one that turns on and off to water your plants at preset times for as long as you want. If your watering routine is an ingrained habit, get a basic timer that you can turn on and off manually at watering times.
- Pipe. Use ABS (black) pipe. Aquapex isn't UV-tolerant, and it needs specialized tools and fittings. PVC pipe is rigid, degrades from sun exposure, and gets brittle and splinters when old. ABS is flexible and easy to work with (especially if you warm it in the sun or with a hairdryer), and it works with many types of connections and attachments.

  When selecting pipe sizes, think about the relationship between flow rate and pressure. The greater the distance covered by the pipe, the lower the water pressure will be. You can compensate for this by choosing pipes of different diameters to maintain, or even increase, the water pressure. If you are covering a large distance or using a lot of emitters, start with larger diameter pipe at the faucet or spigot to maximize your flow, and then use smaller-diameter pipe near the emitters. I have a 1-inch pipe coming off my main spigot and running the full length of the garden. I tapped ¾-inch pipes along the length of the main pipe for the various zones I am watering. Each ¾-inch pipe runs the length of its zone, and ½-inch pipes run from the ¾-inch pipes to each separate bed. Within the beds, I use ¼-inch pipes (coming off the ½-inch pipes) with emitters. This system serves to maintain water pressure throughout the garden despite the wide distribution area and multiple "openings" in the pipes. You may not need such an elaborate setup, but it's a good example.
- Connectors. T and elbow joints will allow you to get the pipe from the faucet to the area that needs watering, as well as to place lengths of pipe above (or next to, or whatever configuration you need) the plants.
- Emitters. Emitters are the prima donnas of an irrigation system; they are the attachments that deliver the

"Soaker hoses" (with holes that emit water) can work with pressurized systems and are most effective on level ground.

water to your plants. You'll be amazed at how many different types there are.

This pressure-watering system from a hose bibb or faucet is intended for the small-yard gardener, but it is equally applicable to any deck or patio area with pots or planters that need watering. I suggest that, before proceeding with the following steps, you take your materials list to the hardware store to familiarize yourself with the options for each of the items you'll need. Helpful hint: Use your cell phone or a digital camera to take photos of the types of items that may apply to your situation. If you need to buy materials during the course of your project, take photos of your garden layout and system to the store with you. I've done this and have referred to the photos countless times.

Materials:
- ☐ Water spigot
- ☐ Female end
- ☐ Screen washer
- ☐ Backflow preventer
- ☐ Water timer
- ☐ ABS pipe
- ☐ Connectors (i.e., T and elbow joints)
- ☐ Emitters
- ☐ Garden clipper or heavy-duty scissors
- ☐ Pressure gauge
- ☐ Pressure reducer (if necessary)

Step 1: Locate/decide on the hose bibb or spigot that you'll be using as the main water source.

Step 2: Measure the distance that a length of ABS pipe will travel from that faucet to your garden area. Use as direct a line as you can, but try to make it an unobtrusive path so that you're not always stepping on the hose or tripping over it.

Step 3: Study the area that you want to water, taking note of the plants or pots, their sizes, and their water needs, to decide what type of and how many emitters you'll need. If many of your plants can handle overhead watering and are in an area that can get wet, misters or sprayers are easy because you need fewer of them. However, some plants can grow mold or fungus from having wet leaves, so you may want soil-level water, which means more emitters but less water wasted.

Step 4: Determine the water pressure from your chosen faucet. Knowing your water pressure will help you decide which emitters to get and how long each watering session should be to deliver the right amount of water to your growing area. Try to borrow a pressure gauge from a neighbor or friendly plumber, or you can buy one (mine was around $6). Some utility companies can tell you the pressure of your house water at the gauge; this may or may not be the pressure at the spigot, but it's a place to start.

TOP TO BOTTOM: ABS pipe in three different diameters, two types of hose clamps, T connector, and various spray heads/emitters.

TOP (LEFT TO RIGHT): inline valve, plug for patching a hole, and elbow connector; MIDDLE (LEFT TO RIGHT): T connector, connector for straight runs, and nail-in tubing staple; BOTTOM: figure-8 for sealing end of pipe.

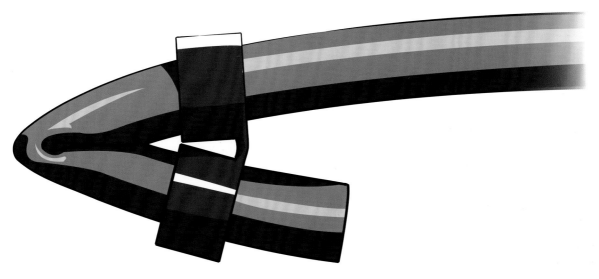

A figure-8 clamp is used to close off the pipe at the end of a run to maintain pressure.

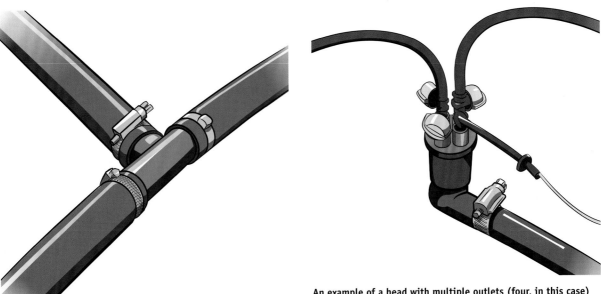

A T connector connects three pieces of pipe and is secured in place by a hose clamp at each joint.

An example of a head with multiple outlets (four, in this case) to which you can attach emitters or sprayers to run to individual plants or sections of the garden.

Get creative! An overhead sprayer is attached to the ABS pipe with an elbow joint and zip-tied to a support to hang above the garden.

Step 5: The fun part: take a trip to the hardware store to gather your materials.

Step 6: The second fun part, in my opinion: go home and set everything up. Cut the ABS pipe with regular garden clippers or heavy-duty scissors (such as poultry shears) and attach the connectors to run the pipe to your garden area. Remember to use the larger diameter pipe closer to the faucet end, decreasing the size, if need be, as you get closer to the plants themselves.

Step 7: Before attaching the emitters to the pipes, run water through the system once to clear out any dirt or debris that may be in the line. Emitters are often pressure-compensating, and even little specks of dirt or sand can affect the flow. Because the flow is light and sometimes tricky to see and gauge, you may not know that you have a problem until the plant begins to show signs of stress or dies.

An example from the author's garden: one pipe comes off the main line to an upright spray head, another snakes up a pole to fill a birdbath, and a third runs to a different section of the bed.

In another example of creative use of her irrigation system, the author clamped a hose with a drip flag emitter to the edge of a birdbath to ensure that the birds would always have water.

Step 8: Follow the instructions that come with your emitters to attach them to the pipes near the plants.

Step 9: When the entire system is assembled, turn it on and let it run for a few minutes. Look carefully at each joint to make sure that there aren't any leaks. Look at the places where water is supposed to be coming out to make sure that it is. Troubleshoot as needed.

Step 10: Once everything looks good, set your timer and pat yourself on the back.

A length of pipe with a simple emitter runs to the base of a potted plant.

Gravity-Fed Watering Systems

In Project 10 in this section, we looked at how to set up an irrigation system. Depending on the size of your growing area, that system may be too big or too elaborate for your needs. A gravity-fed system allows you to use collected or reclaimed water from a rain barrel or set up irrigation where there is no water source otherwise available. A gravity system is a simple irrigation method used in many nonindustrialized countries to minimize water use and maximize irrigation efficiency when water is not ample or easily acquired.

Why would you want to do this?

You want to use reclaimed rainwater or create an "automated" watering system where there is no pressurized water source available.

Why wouldn't you want to do this?

As with the irrigation-system project, you don't have enough plants to bother with a watering system, or you are fine with the routine of hand-watering your plants.

Further refinements:

You can modify this project with different materials to better fit your situation. For example, I have seen buckets hanging from T-shaped posts (such as laundry-line posts) instead of resting on top of bases; this takes up less ground space and could be more versatile. The structure that the bucket rests on can do double duty as a trellis for climbing plants.

Learn more about it:

Visit Chapin Living Waters (www.chapin livingwaters .org); click on "Bucket Kit Gardening." Also see the resources listed in Project 10 of this section (Irrigation Systems).

A gravity-fed system made by Robert and Courtney of His and Hers Homesteading (http://hisandhershomesteading.wordpress.com).

In Project 10 of this section, we looked at setting up irrigation for your yard or garden; the same principles and ideas can be applied to smaller-scale deck or balcony watering. In fact, there are special clamps and brackets designed to attach pipes to walls or baseboards so they stay out of the way and run around the perimeter of your area. Hanging baskets and window boxes benefit particularly from these accessories because you can run the water-carrying pipes above the planters and hang gentle sprinkler heads or misters right above the pots themselves.

In this project, you will learn how to build a gravity-fed watering system for a small garden or balcony area that doesn't have a hose or a faucet for hooking up an automated system. A gravity system uses collected rainwater or other water in a bucket or container that is brought from a different location. Special emitter pipe has holes built into it to distribute water to the plants. Friends of ours set up a gravity system at their community garden plot so they can water deeply without doing it by hand. They use their plot's hose spigot to fill the main cistern bucket.

Things to consider when operating a "low-flow" (less than 10 psi) system:

- Be sure that your emitter pipe is not intended for pressurized systems. It should be labeled as "nonpressure-compensating."
- Keep the distances between water sources and garden areas as short as possible.
- Elevate your bucket/water source as high as possible.
- Test the system by starting with a small version to experiment and get the hang of it.

Materials:

- ☐ Bucket or other multiple-gallon water container (such as a rain barrel) with lid or cover
- ☐ $1/2$-inch-diameter ABS pipe with optional in-line valve
- ☐ Drill with bit the same or slightly smaller diameter than the main-feed pipe
- ☐ Piece of nylon pantyhose or knee-high/ankle-high nylon sock
- ☐ Rubber band
- ☐ Ladder or other support with stable base
- ☐ $1/2$-inch-diameter emitter pipe (called non-pressure-compensating drip line)
- ☐ Any necessary connectors for emitter pipe
- ☐ Caulk (if needed)
- ☐ Joint(s) for ABS pipe (if needed)

Step 1: Decide where you want to lay the lines and how much pipe you'll need to go from the water source to your plants. Although this type of system works best with straight runs, it can work with gentle curves. Try to keep the lines as short and direct as you can. Also consider that the bucket will be sitting on top of a ladder or high shelf; the higher you place it, the better the flow (alternatively, you can hang the bucket from a strong plant hook, rafter, or ceiling hook, keeping in mind the weight of water: a 5-gallon bucket of water will weigh more than 40 pounds).

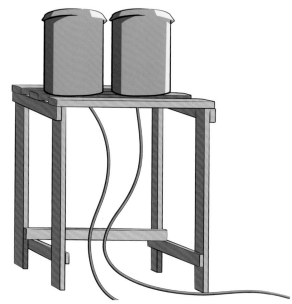

Mount the water source(s) on any stable, elevated support.

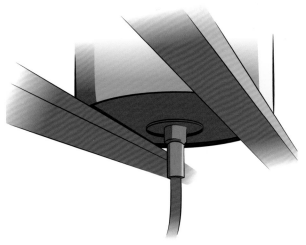

Ensure a tight fit between the pipe and the water container.

Step 2: Drill a hole at the lowest point of the side of or in the bottom of your empty bucket/water container. This hole should be the same size or slightly smaller than the tube (ABS pipe) that will be coming out of it to ensure a tight fit. If it is loose or leaks when water is in the bucket, apply caulk around the juncture of the pipe and bucket.

Step 3: Thread the ABS pipe through the hole and into the bucket, with a few inches sticking in.

Step 4: Double-layer the piece of nylon over the end of the hose that's inside the bucket and secure it with the rubber band; this will serve as the filter. Adjust this end of the hose so that it's situated near the hole with the filter securely attached.

Step 5: Run the ABS pipe to its destination (where you will start the emitter pipe) as directly but unobtrusively as possible. If you need to curve the hose in places, soften it first by laying it out in the sun for fifteen minutes or so or

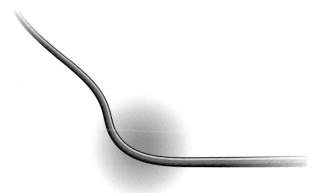

The ABS pipe can handle gentle curves on its way to the plants.

Install joints if you need to form angles; do not bend the line.

by using a blow dryer on the spots that will curve. If you need to make a sharp angle, cut the hose and install an elbow joint; the pipe should not be kinked anywhere.

Step 6: Pour some water into the bucket to flush the line and check for leaks and proper flow.

Step 7: Use connectors to attach the emitter pipe to the ABS pipe as needed so that emitter pipe runs along the bases of the plants. Close off each emitter-pipe end either by folding it back on to itself and securing it with a zip tie or by installing a hose end cap or plug.

Step 8 (optional): If you are using this system with a rain barrel or another type of container that will have water in it regularly, install an in-line valve somewhere on the outflow pipe. Leave the valve open for as long as you'd like to water, and close the valve when the time is up (set a timer to help you remember!).

Step 9: Fill up the bucket from a hose or another bucket of water. Look at your system go!

Run the pipe to the garden to emit water among the plants.

Self-Watering Planters

One drawback to small-scale or container gardening is how fast the planters can dry out. Bear in mind that if you go away on weekends during the summer, all of the good work you put into your patio garden can vanish in just a few short days. Self-watering planters serve as plant-sitters while you are away from home.

Why would you want to do this?
It works well indoors or outdoors, it will save you time watering, and it might save a plant's life if you forget to water.

Why wouldn't you want to do this?
You don't want standing water of any sort, or you prefer to water your plants by hand.

How does this differ from the store-bought version?
A homemade version may not initially look as nice as a manufactured version, but you can get creative and make your planter as fancy as you'd like.

Cost comparison:
The equipment for this project costs less than half of what a prefabricated self-watering planter would cost.

Skills needed:
Basic construction skills.

Learn more about it:
McGee & Stuckey's Bountiful Container (Workman, 2002) by Rose Marie Nichols McGee and Maggie Stuckey; *The Vegetable Gardener's Container Bible* (Storey, 2011) by Edward C. Smith; *Grow Great Grub* (Clarkson Potter, 2010) by Gayla Trail.

Any stackable plastic containers can be used for this project.

I have a tendency to forget to water my potted plants, so I swear by automated watering systems. Many decks or balconies don't have the spigot or pressurized water source needed for automated watering, and in these cases, self-watering containers are a godsend. This project shows you how to build a quick and easy self-watering container in a size that fits your space. It can go anywhere, indoors or out, because the bottom container holds all of the water. Once you know how to make it, you'll start noticing tons of containers in various sizes and styles (it's amazing how many things can be stacked) that you can turn into planters. Warning: this project can become addictive.

Materials:

- ☐ **Two stackable plastic containers (size depends on how large of a planter you want)**
- ☐ **¾-inch or 1-inch PVC pipe (in a length that matches the height of the stack of containers)**
- ☐ **Mesh basket**
- ☐ **Drill with ¼-inch bit**
- ☐ **Utility knife**
- ☐ **Soil**

Step 1: Decide which bin will be inside the other, and mark circles on the base of the inner bin for the mesh basket and PVC pipe near the corner. The holes should be slightly smaller than the diameters of the pipe and basket, respectively, so that the pipe and basket sit snugly in their holes.

Step 2: Cut the bottom of the PVC pipe at an angle.

Step 3: Cut out the circles in the inner container.

Step 4: Flip the inner container over and drill small holes all over the bottom for drainage.

Step 5: Stack the containers and, on the outside of the outer container, mark the level of the bottom of the inner container.

Mark the outer container where the inner container sits.

Cut an overflow hole under the mark you made.

Step 6: Unstack the containers and cut a hole at the mark you made; this is an overflow hole to prevent your soil medium from sitting and soaking in standing water.

Step 7: Stack the two containers and then put the PVC pipe and the strainer into their respective holes in the inner container.

Step 8: Holding the PVC pipe vertically, fill the upper container with soil, tamping it gently. Make sure that the strainer stays in place and is filled with soil.

Step 9: Plant the container and water. Water from the top initially to make sure there is good soil-to-root contact. Once the plants are established, water into the pipe until water comes out of the overflow hole or until you can feel the water level with a finger tucked into the overflow hole. Check the water level weekly to keep the supply high enough for the plants' roots to wick up.

The finished planter, filled with soil and with plants growing inside.

The PVC pipe should extend above the top of the inner container.

Top view of Step 7; the strainer will sit in the large center hole.

Community Gardening

Community gardens offer urban dwellers without gardening space the opportunity to grow their own food in a communal setup. If you're interested in community gardening and you have the chance to participate, don't pass it up—there are probably many people waiting for a spot. So what are the best plants to grow in the limited space provided?

Why would you want to do this?
You want to grow your own food, but you either don't have any available gardening space or you don't want to devote any of your own space to a garden.

Why wouldn't you want to do this?
You have your own garden, you don't want to commit to tending a garden that's not on your property, or you just don't want to grow things.

Skills needed:
Just willingness and energy!

Learn more about it:
Get involved! Talk to the administrator of your community garden to find out about gardening classes or Master Gardener programs. Walk around your community garden to look at what other people have done. Ask other community gardeners what works—and what doesn't—for them. There is a wonderful children's book called *Seedfolks* (HarperTrophy, 2004), by Paul Fleischman, which tells the story of how a vacant lot became a community garden and how working on the garden united the community and turned a troubled area into an oasis of sorts.

Willingness to put in the required labor is a key factor in community-gardening success.

If you are reading this, chances are that you either want a community garden plot or have recently gotten one and are trying to decide what to grow. One of the great things about community gardens is the first part of the name: *community*. Community gardening is a great way to begin gardening and to learn through both hands-on experience and the collective knowledge of the other gardeners out there. Many people I know have gone from newbies to expert gardeners just from what they've learned, seen, and experienced by participating in community gardens.

Community gardens are wonderful. In our city, there are currently thirty-eight of them that offer 1,200 plots and have approximately 3,000 people working them. Each of our community gardens has a substantial waiting list for plots as they become available.

Most cities with community-gardening programs have guidelines about what you are allowed and not allowed to grow in the community plots. Generally prohibited are prolific or spreading plants, such as mint, comfrey, lemon balm, bronze fennel, or other invasive species, that would take over the entire area. For practical reasons, there are other plants that just don't make sense to grow in an average community garden plot.

I spoke to some community gardeners and got their tips on what to grow in a community plot.

Advice from Community Gardeners

Gardener A: First off, grow the things you really like—or learn to really like the things you grow.

Gardener C: Try to grow things that don't take up a lot of space but can yield a lot, such as peas, beans, or cucumbers. Climbing [plants] are great because they take up minimal square footage and can be trellised to make use of the air space too. Root vegetables and greens are good for the smaller space they take up. Don't try things that require a lot of space (for example, zucchini or corn) for relatively little reward—unless you really love it.

Gardener A: Space considerations do preclude [plants] such as corn, but we've done it occasionally. We like the perennials and things that seed themselves, such as chard—it is great to walk in with a knife and come back out with greens for the skillet.

What Should You Grow?

Easy to Grow	Prolific Growers	High Yield/ Small Footprint	Attractive	Climbers	Like Ample Water	Attract Good Bugs	Expensive to Buy
• Arugula • Beans • Cucumbers • Garlic • Kale • Lettuce • Parsley • Parsnips • Peas • Potatoes • Radishes • Sunflowers • Tomatoes	• Beans • Chard • Cucumbers • Kale • Parsley • Peas • Tomatoes	• Basil • Beans • Cucumbers • Eggplants (Japanese) • Parsley • Peas • Hot peppers • Tomatoes	• Beets • Chard • Eggplant (Japanese) • Kale • Hot Peppers • Sunflowers	• Beans • Cucumbers • Peas • Tomatoes (trellis) • Sunflowers	• Basil • Celery • Lettuce • Melons • Parsley • Tomatoes	• Cucumbers • Cilantro • Dill • Sunflowers	• Basil • Bell peppers • Eggplants (Japanese) • Herbs • Kiwis • Strawberries • Tomatoes

Gardener B: Because space is limited, I think it is worth learning more about vertical gardening and talking with neighbors about trades of seeds or harvest.

Gardener A: I think that economically it makes sense to grow the things that are expensive to buy, such as raspberries, and to buy the things that are cheaper, such as potatoes and onions.

Gardener B: In Portland, community-garden rental is a flat rate with no extra fee for water usage. So while space is limited and I wouldn't do watermelons, I would do smaller thirsty crops such as melons, berries, and tomatoes because I won't get charged for the water. (Check your own garden's policies and fees.)

Gardener C: Put in [plants] that can give you fairly fast turnaround so you don't spend years waiting for your first harvest only to decide to leave the plot just as long-term plants, like asparagus or rhubarb, are maturing.

Gardener A: As far as [plants] that take time to establish, such as asparagus, it depends on how long you plan to have the plot and how much space you want to dedicate to one crop. We have rhubarb and raspberries as anchor plants, but most of the rest of the footage is flexible.

Gardener B: If you live close to your garden, grow things that can be picked fresh each day—herbs, berries, salad greens. If you live farther [away], grow rare versions of things that hold, such as an heirloom potato you might love but would never find in the market, green beans that can hold for several days when ready, or hardy greens such as kale and chard.

Gardener A: Always plant tomatoes because there is nothing like a BLT in August or September—best sandwich of the year!

A community garden in bloom with flowers and greens.

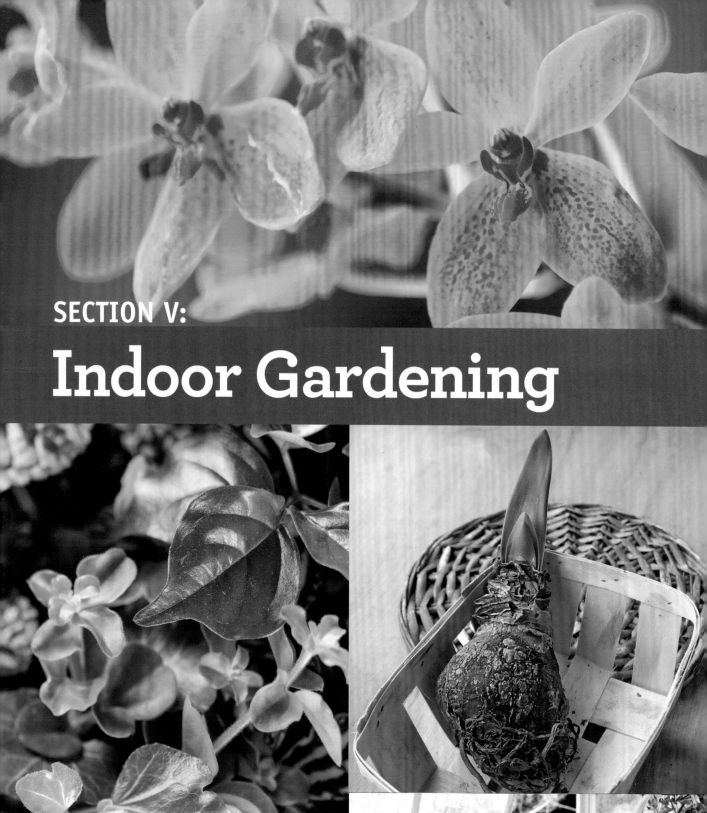

SECTION V:
Indoor Gardening

Windowsill Herb Gardens

Even if your only exposure to the outdoors comes when you exit the front door of your apartment building and set foot on the city sidewalk, you can still have a garden. If you have a windowsill—or even just a window—you can grow delicious additions to your table.

Why would you want to do this?
You really want to try growing your own food, but you have no outdoor gardening space. Plus, herbs are easy to grow and great to have on hand.

Why wouldn't you want to do this?
You don't want to grow plants, you don't cook much, or you don't use herbs in cooking.

How does this differ from store-bought versions?
Many purchased herb plants are heavily fertilized, so they may be lusher than yours, but your homegrown herbs will be more flavorful.

Cost comparison:
It is far more expensive to buy dried herbs than it is to dry some branches or leaves from your own plants.

Learn more about it:
Visit the Windowsill Gardening page on the Harvest to Table website (http://harvesttotable.com/2010/01/window sill_gardening_growing_v/) and check out *The Kitchen Garden Grower's Guide* (BookSurge, 2008), both by Master Gardener Stephen Albert.

Further refinements:
Culinary uses are practically limitless, and you can create plenty of personal and household products from your homegrown herbs.

Certain flowering plants also fare well indoors.

Drainage Tips

If a container does not have drainage holes, put a layer of pea gravel an inch or two thick in the bottom of the planter, place a used dryer sheet over the gravel, and put the soil on top of the dryer sheet. This will prevent your plants' roots from rotting by giving the standing water a reservoir. Be judicious when watering pots that have no drainage; they should only receive water when the soil feels like it is drying out. To be certain, you might want to occasionally dig your fingers down below the top surface to see whether it's damp at all.

Herbs are among the most expensive items in the grocery store, and it always surprises me that anyone who can grow his or her own herbs would buy them. One of the reasons that herbs are so widely used in cooking is they are easily cultivated. Many of them count as weeds in their native areas and have growing traits that follow suit: they can grow in extreme conditions, have large yields, and attract pollinators when grown outdoors. The end result is durable plants that are tasty, edible, and healthy.

Because they are easy to grow and do not need much space, herbs have become very popular indoor plants. Sunny city and suburban windowsills around the world are alive with the popular basil, thyme, rosemary, oregano, and sage, as well as more exotic herbs. Most "aromatic" herbs serve as beneficial companion plants for other edibles, and many can be grown together in a container to mutual benefit.

One of the many things I like about windowsill gardens is how good the plants always look. I think that indoor plants are typically nice and healthy because they get so much care. They are small, it takes no time at all to water and trim them, and you won't have any weeds to deal with. Because mine are right above my kitchen sink, I look at them and give them attention throughout the day, every day. Instead of dumping my half-full water glass down the drain, I give the water to my plants. They get steam when we do the dishes, they bask in morning and midday sun, and they enjoy the ambient temperatures of our most inhabited room. It's a symbiotic relationship: I nurture the plants, and they reward me with good food or flavorful spices or even a soothing salve (from the aloe vera plant, good for cooking burns), as well as pretty decor. If you can find several containers that look nice lined up side by side, you'll have the makings of a lovely windowsill garden, a little oasis of your own.

Materials:
☐ Small planters or containers
☐ Drip trays
☐ Potting soil
☐ Small rocks (to help with drainage)
☐ Herb plant starts
☐ Small kitchen shears

Mix and match your plants and your planters for an attractive and functional windowsill garden.

A sunny windowsill is the perfect spot for indoor herbs.

Step 1: Find three to five containers that fit on your windowsill. This is where your creativity comes in: use tin tea containers of different colors. Use bright ceramic mugs, either a matched set or several in different colors and sizes. Get a series of small to medium terra-cotta pots and a large drip tray (washed Styrofoam meat trays make good drip trays) to set them in, and paint the rims with patterns or your favorite quotations. Line up three or four interesting bowls. Collect the cans from a few weeks' worth of soup, tear off the labels, and use clear packing tape to seal interesting photos to the outside of each can. Alternatively, you can use one long, thin container that fits the length of your windowsill, such as a piece of copper or aluminum gutter with ends attached. Whatever you do, don't spend a bunch of money—you can find plenty of options around the house.

Step 2: Put drainage holes in the bottom(s) of your container(s) (ceramic containers are particularly challenging, so unless you know how to drill holes in them—with a special drill bit and a ton of patience—skip this step) and place the containers on a drip tray. Put a layer of rocks on the drip tray to help keep the root zone from getting too soggy and to create more humidity around the plants.

Step 3: Fill all of the containers three-quarters full of potting soil.

Step 4: Plant small herb plants in each container. If the container is large enough, you can plant more than

Windowsill Herb Gardens **207**

Cacti are very forgiving indoor plants if you want to start off with something very easy.

one plant in it, either several of the same herb or a few different herbs. Good herbs for small containers are thyme, rosemary, parsley, and small-leafed basil. You can also try a small hot-pepper plant, such as cayenne or ornamental.

Step 5: Top the containers with soil at the plant bases, firming the soil around the edges. Give the pot an initial soaking, both to settle the soil and to make certain there is good root-to-soil contact.

Step 6: Water the plants carefully—you want them to dry out, but not become parched, between waterings (see Drainage Tips sidebar for advice on planters without drainage).

Step 7: Initially, trim your herbs for kitchen use lightly. Once the plants are established, you can trim them more regularly—this will encourage growth, which in turn encourages root growth (more roots take up more water, which helps prevent waterlogging).

Plants for a Basic Indoor Herb Garden

This is just a partial list of herbs that you can grow easily in your garden, yielding plenty to harvest fresh and to dry for use when the plants are dormant. You may never buy basic herbs from the grocery store again!

Basil: This leafy green plant is a perennial that most of us treat as an annual. It is cheap and easy to start from seed, and it likes a lot of heat. Basil prefers full sun but can actually sunburn in hot climates. It is not one of the more drought-tolerant herbs; give it water regularly to ensure plentiful bright-green growth. Too much water and not enough light will yield long, leggy stems and smaller leaves. There are many different varieties beyond the basic flat-leafed green type. Try purple leaves or lemon-flavored basil for a change of pace. Some basil plants have leaves as large as lettuce leaves; others have plentiful, tiny, spicy leaves.

Chives: Chives are relatives of the onion and are incredibly versatile. They are a culinary delicacy of sorts because they don't store well and they lose their pungency when dried. Chives grow from a cluster of bulbs and will readily grow back year after year (and month after month), multiplying their bulbs annually. They should be dug up and divided every three to five years. The chive "leaves" can be trimmed in small handfuls and snipped fresh over many different foods, or they can be chopped and cooked for a gentler, more nuanced flavor. Let your chives flower—the pretty pincushion flowers are edible. You can take individual blossoms from the round flower head and add them to a salad or on top of pasta. Each flower gives a tiny burst of chive flavor and adds color.

Cilantro: This annual herb is easy to start from seed but is hard to grow as the green leaf we all know and love in salsa. I start a pinch of seeds (at least six) in a single pot and give it good light (but not too much full sun) and lots of water for ample leaf growth. If it dries out, it will bolt, which means that it is under stress (*"Aaaack! Life in peril! Must reproduce before I die!"*) and will send up a flower/seed stalk. The silver lining when cilantro bolts: the seeds that form are coriander, another spice. If you let it bolt, collect the seeds when the tops are brown and dry. Grind some of the coriander in a coffee grinder, and you've got a staple seasoning in Indian and Cajun cooking; plant the rest of the seeds for cilantro next year.

When your plants are nearby, you're very likely to give them TLC.

Oregano: I love oregano and use it in much of our cooking, but I get irritated that it overruns less aggressive herbs. There's only so much oregano that one can use fresh, and I still have plenty left to dry. This is a hearty plant that can be found on the dry, rocky slopes

Oregano is a prolific grower whether outdoors or indoors.

The easy-to-recognize parsley is a biennial plant.

of the Mediterranean; there are Greek and Italian varieties and lots of choices in leaf and flower color and size.

Parsley: This nutrient-packed green normally grows for two years (a biennial), putting on the leafy growth in the first year for maximum photosynthesis to nourish the root. In the second year, the root uses that stored energy to send up a few leaves and a prominent flower head that then yields the seeds for next year's new plants.

Rosemary: People who enjoy Greek food recognize the importance of rosemary in Mediterranean cooking. This is an aromatic herb with both upright and prostrate varieties. Rosemary tolerates extreme heat and drought better than cool, moist conditions.

Rosemary is flavorful and aromatic.

Dry Your Herbs

Long method: Hang branches or leaves from a rafter in a warm, dark spot with plenty of air circulation, such as an attic in the summer or a warm furnace room. Leave them hanging until the leaves crumble to the touch.

Quick method: Spread leaves in a single layer on a paper towel. Microwave on high for three to four minutes or until the leaves crumble to the touch.

Store dried herbs in sealed containers in a dark place away from light and heat. Try to dry and store only what you'll use in a few weeks' time because the oils that give them their flavors will degrade.

Sage: The *Salvia* genus is one of the largest in the plant world, and there are so many varieties to choose from. On the edible varieties, you'll find leaves in different sizes and colors—from green to bluish green to purple to variegated— as well as flowers in almost all shades of the rainbow. The standard culinary sage is a good, hearty plant and is yet another indispensable cooking herb. Some sage flowers are edible and make lovely salad additions.

Thyme: Thyme is one of the most popular herbs from my herb garden. Thyme likes a warm area and needs good drainage. It will overwinter in cooler climates as long as it is not waterlogged. Thyme is a low grower and can go in a pot with more upright herbs or at the edge of a planter, where it will creep and cascade down the sides. There are many different varieties of this herb, from regular culinary thyme to lemon, lime, variegated, and other types. It is indispensable in cooking.

The different varieties of thyme have many uses in the kitchen.

Setting Up Artificial Lighting

You don't have to have a greenhouse or an industrial-strength grow lamp to grow plants inside. You can keep your houseplants healthier and blooming longer than they would with just normal, ambient house lighting, or you can keep them growing off-season with the proper use of artificial lighting.

Why would you want to do this?
You have room to grow plants indoors, and you want to try to extend their growing season and maximize the possibilities of what you can grow.

Why wouldn't you want to do this?
You are concerned about using too much electricity or increasing your electricity bill, or you live in an area with plenty of sunlight hours.

Is there an easier way?
You can buy prefabricated grow-light stands, but they are expensive. You can simplify this project by using (or building) simple bookshelves with securely attached cup hooks on the undersides of the shelves. Attaching S hooks to the lights and employing lightweight chains makes it possible to raise and lower the lights depending on growth and desired light intensity.

Cost comparison:
You can make any of the versions of the grow-light stand in this chapter for about half, or less than half, of the price of a prefabricated stand.

Skills needed:
Basic construction skills.

Learn more about it:
Gardening in Your Greenhouse (Stackpole Books, 1998) by Mark Freeman describes lighting options and includes illustrations; *The Complete Book of the Greenhouse* (Ward Lock, 5th ed., 1996) by Ian G. Walls has a brief chapter on light and plant growth; *Greenhouse Gardener's Companion*, Revised, (Fulcrum, 2000) by Shane Smith can't be beat for thorough explanations of artificial lighting and its role in plant growth.

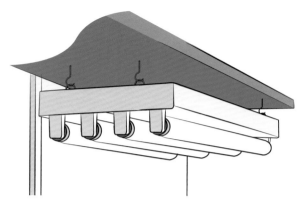
You can hang the light fixture flush or use chain to lower it.

In Version Three, the light hangs under the sawhorse crosspiece.

Obviously, we are no longer dependent exclusively on the natural seasons to obtain fresh produce. Supermarkets, industrial manufacturing processes, and long-distance shipping of specialties have almost rendered seasonal eating moot. That is not necessarily always a good thing—we now know that it is healthier to eat less processed foods, eat seasonally, and grow as much of our own food as we can. However, technology can be our friend by helping us maximize our abilities to grow and to extend, rather than disregard, the seasons.

Using artificial light to grow plants is almost as time-honored as electricity itself, and there's such a variety of affordable light sources available. Incandescent lights work to start seeds and plants in a pinch, but over the years, growers have learned that different types of light emit different ranges on the light spectrum, as well as different heat output.

I had a hand-me-down high-powered metal halide grow light, and only when it was stolen out of my greenhouse did I find out how many varieties of grow lights exist and how much they cost. Professional grow lights are insanely expensive (and cause your electricity bill to skyrocket). You can just as easily use a regular light fixture and, with the benefit of the warm environment of your home, start and grow plants for a fraction of the cost of using a grow light.

Professional grow lights can generate an ideal light spectrum, but they (and incandescent bulbs) generate too much heat. I have had many flats of fragile seedlings actually baked from being too close to the light. The spectrum of fluorescent bulbs is not as complete as that of other lights, but because they hardly generate heat, it is fine—and important—to put fluorescent lights close to new plants. The bulbs should be placed no more than 4 inches from the growing medium (prior to sprouting) or the top of the highest leaf to provide the proper amount of light for growth.

Because I packed up and moved at least twice a year during my college years, I appreciate objects that can fold flat; to gardeners, seasonal items that can be dismantled and stored in small places are invaluable. The following DIY artificial-light fixtures compact to not much more than the size of the light itself.

Materials for all projects:
- [ ] Standard work light fixture (made for tube-style fluorescent bulbs) for two (minimum) or four (ideal) bulbs
- [ ] Fluorescent plant or aquarium bulbs that fit your fixture (although these are sometimes labeled as "grow lights," these fluorescent lights do not generate the heat that the high-powered professional grow lights do; rather, these are fluorescent bulbs with a spectrum of light specifically for growing plants or lighting aquariums)
- [ ] Small link (1-inch or less) chains, length determined by your site
- [ ] S hooks
- [ ] Optional: screw eyes

The first step for all versions of this project is to assemble the work light with the bulbs installed. These types of lights usually have openings or hardware on top for hanging. Instead of using the included hardware, insert S hooks (or, better yet, screw eyes and then S hooks) into the top openings. You want at least two hangers, and they should be evenly spaced down the length of the light. Whichever version of this project you use, you must turn the light off for at least eight hours a day.

Version One

The beauty of this system is that, as the plants grow, you can raise your light in small increments by either moving the S hooks up the chain or shortening the chain at the ceiling hooks.

Additional materials:
☐ Ceiling hooks or screw eyes, one per hanger

Step 1: Securely install the ceiling hooks or screw eyes in your ceiling above where your plants or flats will go. Space them to match up with the hangers in your light.

Step 2: Hang an appropriate length of chain from each of the hooks/eyes, using S hooks if necessary, so that your light will hang at the desired distance from your plants.

Step 3: Hang the light fixture from the chain, adjusting the distance from the plants or the flats if necessary.

Version Two

Additional materials:
☐ Brackets, one per hanger

Note: The distance of the light from the plants can be adjusted here, too, but this version uses shelf brackets that are mounted to a wall. The arm of the bracket that protrudes from the wall must be as long as at least half the width of the light fixture so that the fixture will hang away from the wall.

Step 1: Install the brackets securely in the wall above where your plants or flats will go. Place them the same distance apart as the hangers in your light, and make sure that they are level with one another.

Step 2: Hang an appropriate length of chain from each bracket, using S hooks if necessary, so that your light will hang at the desired distance from your plants.

Step 3: Hang the light fixture from the chain, adjusting the distance from the plants or flats if necessary.

Version Three

Additional materials:
☐ Two sawhorse brackets
☐ One length of 2x4, 6 to 8 inches longer than the length of your work light
☐ Four lengths of 2x4, slightly longer than the desired maximum height for your grow light
☐ Ceiling hooks or screw eyes, one per hanger

Note: While this version is not as pretty (or as unobtrusive) as the previous versions, it has the advantage of needing less chain and being more versatile for location. It is ideal for cool-temperature fluorescent lights because the distance between plant and light is minimized. It can go on a tabletop, under a table, or even in a closet. And when not in use, the entire thing can be taken apart and easily stored.

Step 1: Assemble the sawhorse with the 2x4s according to the instructions that come with the brackets.

Step 2: Install the ceiling hooks or screw eyes on the underside of the crosspiece to match up with the hangers on your light.

Step 3: Hang an appropriate length of chain from each of the hooks/eyes, using S hooks if necessary, so that your light will hang at the desired distance from your plants.

Step 4: Hang the light fixture from the chain, adjusting the distance from the plants or the flats if necessary.

Version Three is best for plants that need close proximity to light.

Creating Miniature Topiaries

Many people find the art of topiary beautiful and intriguing, but it is also labor-intensive and detrimental to the growth of the plants (preventing them from reaching their full growth). Why not practice this craft on your own windowsill? You'll have freshly snipped herbs to use for cooking at the same time.

Why would you want to do this?
It's easy and fun, and the result makes either a great gift or a useful and manageable indoor herb or flower garden.

Why wouldn't you want to do this?
You can't grow plants inside your house, or you dislike the look of topiary.

Skills needed:
Green thumb with indoor plants.

Learn more about it:
The Complete Book of Topiary (Workman, 1988) by Barbara Gallup and Deborah Reich gives great ideas for plants that can be made into topiary, as well as design ideas and drawings that show a step-by-step process of creating one. For a look at real topiaries from gardens around the globe, check out *A Practical Guide to Topiary* (Anness, 2012) by Jenny Hendy, which has hundreds of photos of different styles.

Topiary adds a bit of luxury to your home—your own oasis in everyday life.

The topiary shapes of the gardens at the Palace of Versailles, the pleasure grounds of King Louis XIV, are often held up as an ideal of garden perfection—harmony, pleasing aesthetics, peace, and structure. The idea that we can actually control our gardens—the vision of a picture-perfect, not-a-leaf-out-of-place landscape (maintained by someone else)—is a daydream that can settle the mind of a harried householder. When you try your hand at small-scale topiary, you can create your own sense of place and come home to your very own miniature Versailles—easily pruned and shaped, and even useful in helping you season that coq au vin in the Crock-Pot!

Like bonsai, the ancient Asian art of growing and training miniature trees and shrubs, this form of topiary uses small-leafed plants to create miniature works of art similar to those seen in full-size gardens staffed by professional crews. It is easy to do, cheap, and fun, and you can use the end products as gifts for friends or relatives or to enhance your culinary creations. For those that you keep for yourself, I find that it takes only a few minutes each week to water and trim them to maintain their shapes. Looking at your topiaries can serve as a mini-vacation to a land of formal gardens, complete with the satisfying aromas of lavender from France, rosemary from Italy, or thyme from Greece.

Materials:
- ☐ Tiny potted herbs or flowering shrubs
- ☐ Small, sharp, pointed scissors or shears
- ☐ Wire and wire cutters, optional

Step 1:

Option 1: Start your own plants by rooting cuttings in water. This can give you nice, straight stems to begin your topiaries, which works well for the standard ("lollipop") shape. The benefits of starting your own cuttings are that you pick the stem shape and size from the plants' infancy, and it's quite a bit cheaper than buying nursery plants; the downside is the time it takes for the cuttings to grow large enough to shape (about a 5- to-7-inch stem with growth).

A heart shape on a wire frame.

218 Urban Farm Projects

Option 2: Buy plants in 4-inch pots, which should have good-sized stems or ample creeping growth. This jump-starts your topiary, but it can be expensive, and it limits you to the plants that the nursery has on hand. Consider your choices based on the shapes you hope to achieve.

Step 2: Repot the plants into a larger pot to encourage new growth and vigor.

Step 3: Decide what shape you want the final topiary to be. Your options are limited only by your imagination and the growth patterns of the plant. You can tie long, straight stems onto wire frames, and you can twist vines carefully around circular or heart-shaped mini-trellises. If you have a fuller, shrublike plant, you can use a three-dimensional wire frame in whatever shape you'd like and prune the errant growth back to the frame. If the plant has a strong central stem, you can tie it gently to a stake and allow it to grow upright into a "trunk." Once it is tall enough, you can prune the main growth to make a full-topped standard.

Step 4: If you are using a wire structure, whether two-dimensional (such as a heart) or three-dimensional (such as a cube), cut and bend the wire (I use hangers) into the desired shape and size, leaving prongs so that you can push the frame into the soil at the base of the plant (2-D) or centered over the plant (3-D). If you're using a three-dimensional frame and your plant has a lot of growth, you can tuck the plant into the center of the frame.

A double lollipop on a nice, straight stem.

Favorite Plants for Small Topiaries

HERBS	ORNAMENTALS
Lavender	Small-leaved fuchsia
German chamomile	Cardinal climber or other small-leaved vines
Lemon verbena	Scented geranium (*Pelargonium*)
Rosemary, either upright or trailing	Dwarf/miniature rosebush
Thyme	Dwarf conifer (check with bonsai growers)
Small-leaved basil	Juniper
Caper bush	

Use your imagination and the growth pattern of your plant to give you ideas for the final shape.

Step 5: Water regularly and give the plant lots of natural or strong, artificial light to promote short, sturdy growth. This is the hardest step for me, because it also requires patience. You can't force a plant to grow faster than it will. If you try to, the result is often long, spindly, weak growth that is more prone to breaking and disease, and it will not form the thick, leafy stems you want for a full, shapely topiary. You can also fertilize the plant with plant food product such as Miracle-Gro, but again, be moderate. Too much chemical fertilizer can actually burn the roots, either killing the plant, or promoting long, spindly growth that is not conducive to strong topiary training and trimming.

Step 6: As the plant grows, trim it occasionally. If it is in a three-dimensional shape, just trim off the growth that has extended outside the wire frame. If it is a twining plant or one you are training in a linear shape, gently twist new growth onto the wire trellis and trim off errant growth. A good way to promote bushy growth, especially on fuchsias and other ornamentals, is to pinch out the growth tips in the middle of a pair of branching leaves, like removing the buds from basil. Doing this with every branching pair results in exponential growth as well as a bushier habit. On multi-stemmed, fine-leaved plants like thyme or rosemary, just pretend your clipper is a miniature hedge trimmer, and snip around the perimeter to shape it.

Other Ideas

Try starting three fuchsias, each with a different blossom color, and braid them together in one pot for a tricolored standard. Create a small espalier by making a wire frame to grow the main branches along, and trim those to the form. Or pull longer side branches down to soil level, pinning down with a stake, and repeating it as they root. You can grow a little row of vertical branches like a miniature windowsill hedge to block the view of your neighbor's trash cans. How about three different flavors of thyme with braided "trunks" and a multicolored crown? Or a variegated fuchsia twisted with a solid green-leafed one? Or a small vine twining in the shape of the first letter of someone's name? The choices are limited only by your creativity.

Step 7: Repeat steps 5 and 6 until the plant reaches a size and shape that you are happy with. At this point, it is complete and, with a ribbon tied around it, or placed or replanted in a pretty, decorative container, makes a perfect gift. Alternatively, keep it on your kitchen windowsill to give you frequent, short-term doses of nirvana from the sense of control you have over this one small object in your house. Use the clippings from herbs in your cooking, or leave them out to dry for storing in spice jars. Tranquility, control, beauty, and culinary satisfaction, all wrapped up in one small package!

When starting with ample growth, trim and use supports to train your plant into your desired shape.

Insert a wire frame at the base of a new plant to encourage growth up and around the frame.

Resources

In addition to the resources I've included in the "Learn more about it" section that precedes most of the projects, I've used and I recommend the following:

BOOKS

Bremness, Lesley. *Complete Book of Herbs, The.* New York: Penguin Studio, 1994.

Bumgarner, Marlene Anne. *New Book of Whole Grains, The.* New York: St. Martin's Griffin, 1997.

Burton, Robert. *National Audubon Society North American Birdfeeder Handbook.* New York: DK, 1995.

Cobleigh, Rolfe. *Handy Farm Devices and How to Make Them.* New York: Orange Judd Company, 1909.

Coleman, Eliot. *The New Organic Grower.* White River Junction, VT: Chelsea Green Publishing, 1995.

Coney, Norma. *The Complete Soapmaker.* New York: Sterling, 1997.

Flottum, Kim. *The Backyard Beekeeper.* Beverly, MA: Quarry Books, 2005.

Head, William. *Gardening Under Cover.* Seattle, WA: Sasquatch Books, 1989.

Hartung, Tammi. *Growing 101 Herbs that Heal.* North Adams, MA: Storey Publishing, 2000.

Ody, Penelope. *The Complete Medicinal Herbal.* New York: DK, 1993.

Pangman, Judy. *Chicken Coops: 45 Building Plans for Housing Your Flock.* North Adams, MA: Storey Publishing, 2000.

Papazian, Charlie. *The New Complete Joy of Homebrewing, rev. ed.* New York: Avon Books, 1991.

Rehberg, Linda, and Lois Conway. *Bread Machine Magic.* New York: St. Martin's Griffin, 1992.

Roberts, Michael. *Making Mobile Hen Houses.* Devon, England: Gold Cockerel Books, 2005.

Ruhlman, Michael. *Ratio: The Simple Codes Behind the Craft of Everyday Cooking, reprint ed.* New York: Scribner, 2010.

Smith, Cheryl K. *Goat Health Care.* Cheshire, OR: karmadillo Press, 2009.

Weaver, Sue. *Chickens.* Irvine, CA: I-5 Press, 2005.

Weaver, Sue. *Goats.* Irvine, CA: I-5 Press, 2006.

MAGAZINES

We don't subscribe to many topic-specific magazines, and we hold the few that we get to high standards. These are our favorites.

Brew Your Own
www.byo.com
An informative and entertaining resource for home brewers of all experience levels.

Cook's Illustrated
www.cooksillustrated.com
Our friend Jeff (of the "Sand Wedge" bread recipe) swears by this one, and it has helped us in our cooking endeavors many times. It doesn't have any advertising, which we especially love.

Fine Cooking
www.finecooking.com
Beautifully illustrated and packed with recipes, advice, and articles for the foodie, this is my husband's lay-awake-late-at-night read.

Grit
www.grit.com
A rural-lifestyle magazine covering all topics related to living in the country.

Growing For Market
www.growingformarket.com
A journal-type trade publication for those who grow and produce their own food as a business, large or small.

Small Farm Canada
www.smallfarmcanada.ca
A bimonthly whose mission is to promote "small-scale farming as a legitimate and viable endeavour."

Urban Farm
www.hobbyfarms.com/urban-farm
An I-5 Publication known for its beautiful photography and wealth of DIY advice for urban and suburban dwellers.

WEBSITES/RETAIL RESOURCES

Cosmetics Info
www.cosmeticsinfo.org
A resource for learning about ingredients and safety regarding makeup and other personal-care products.

Fias Co Farm
www.fiascofarm.com
A resource for herbal and holistic practices, goat keeping, and making your own dairy products.

Gardener's Supply Company
www.gardeners.com
One of the retailers from whom I purchase many of my supplies, and I look forward to receiving their print catalogs as well.

Growers Supply Company
www.growerssupply.com
Another retailer I often count on for supplies and a good catalog.

Lee Valley Tools
www.leevalley.com
And a third retailer and catalog supplier that has proven very handy.

Murray McMurry Hatchery
www.mcmurrayhatchery.com
A supplier of poultry, poultry equipment, and advice on raising poultry, all geared toward the small farmer.

Nichols Garden Nursery
www.NicholsGardenNursery.com
A supplier of seeds and herbs for the "gardener cook" whose website also offers a blog and a variety of "garden to table" recipes.

Photo Credits

All illustrations by Tom Kimball

Cover: *front* (clockwise from top): RDPixelShop/Flickr, Zigzag Mountain Art/Shutterstock, Olena Mykhaylova/Shutterstock, Alison Hancock/Shutterstock, Christopher Elwell/Shutterstock *back:* c.byatt-norman/Shutterstock *author photo:* Roger Wood

title page: federicofoto/Shutterstock **page 4:** (left) Mircea BEZERGHEANU/Shutterstock, (top right) Alison Hancock/Shutterstock, (bottom right) Heike Rau/Shutterstock **page 5:** Cynthia Kidwell/Shutterstock **pages 6-7:** littleny/Shutterstock.com **page 8:** Ingrid Balabanova/Shutterstock

SECTION I

page 10 (clockwise from top): Sandra Besic/Shutterstock, Ingrid Balabanova/Shutterstock, Heather McCann/Shutterstock, GekaSkr/Shutterstock **page 11 (clockwise from top left):** Dallas Events Inc/Shutterstock, martiapunts/Shutterstock, Heike Rau/Shutterstock **page 12:** Theresa Carpenter/Flickr **page 13:** chirapbogdan/Shutterstock **page 14:** Warren Layton/Flickr **page 15:** (top) Warren Layton/Flickr (bottom) Mageon/Shutterstock page 16: Aleksandra Duda/Shutterstock **page 17:** Angelo Giampiccolo/Shutterstock **page 18:** Brian Chase/Shutterstock **page 19:** (top left) Olga Nayashkova/Shutterstock (top right) Hanan Cohen/Flickr (bottom left) Olga Nayashkova/Shutterstock **page 20:** High Voltage/Flickr **page 21:** mipstudio/Shutterstock **page 22:** Berents SS **page 23:** aquariagirl1970/Shutterstock **page 24:** (top) Rebecca Siegel/Flickr (bottom) Tobik/Shutterstock **page 25:** (top) ilovebutter/Flickr (bottom) O.Bellini/Shutterstock **page 26:** Rebecca Siegel/Flickr **page 27:** Celeste Lindell/Flickr **page 28:** Madlen/Shutterstock **page 29:** Gayvoronskaya_Yana/Shutterstock **page 30:** Rachel Tayse/Flickr **page 31:** (top) Smit/Shutterstock (bottom) Bratwustle/Shutterstock **page 32:** Jessica Spengler/Flickr **page 33:** (top) Madlen/Shutterstock (bottom) D.Shashikant/Shutterstock **page 34:** nito/Shutterstock **page 35:** Nuiiko/Shutterstock **page 36:** Jeff Keacher/Flickr **page 38:** (top) http://jronaldlee.com/Flickr (bottom) clkohan/Flickr **page 39:** Joanna Poe/Flickr **page 40:** (top) Like_the_Grand_Canyon/Flickr (bottom) Yoshihide Nomura/Flickr **page 41:** cheeseslave/Flickr **page 42:** Vitaly Korovin/Shutterstock **page 43:** udra11/Shutterstock **page 44:** Robert S. Donovan/Flickr **page 45:** Andrew Chellinsky/Flickr **page 46:** Olga Miltsova/Shutterstock **page 47:** Luiz Rocha/Shutterstock **page 48:** oznuroz/Shutterstock **page 49:** Katherine/Flickr page 50: tehcheesiong/Shutterstock **page 51:** 3523studio/Shutterstock **page 53:** (top) Käfer photo/Shutterstock (bottom) Nitr/Shutterstock **page 54:** Evgeny Karandaev/Shutterstock **page 55:** Peter Radacsi/Shutterstock **page 56:** Ty Konzak/Flickr **page 57:** (top left) Simon Law/Flickr (top right) mexrix/Shutterstock (bottom) Diana Taliun/Shutterstock **page 58:** tomer turjeman/Shutterstock **page 59:** maxfeld/Shutterstock **page 60:** Fotofermer/Shutterstock **page 61:** Chamille White/Shutterstock **page 62:** (top) Warren Price Photography (bottom) Zigzag Mountain Art/Shutterstock **page 64:** imaller-

gic/Flickr **page 65:** JoePhoto/Flickr **page 66:** draconus/Shutterstock **page 67:** Ronald Sumners/Shutterstock **page 68:** Krzysztof Lis/Flickr **page 69:** Milan Tesar/Shutterstock **page 70:** mashe/Shutterstock **page 71:** Tata/Shutterstock **page 74:** Valentina R./Shutterstock **page 75:** Alexandralaw1977/Shutterstock **page 76** (top) B.D.'s world/Flickr (bottom) nola.agent/Flickr

SECTION II

page 78 (clockwise from top): dedi57/Shutterstock, Layland Masuda/Shutterstock, Steve Wood/Shutterstock, Sk Elena/Shutterstock **page 79** (clockwise from top left): Andreja Donko/Shutterstock, mythja/Shutterstock, Olena Mykhaylova/Shutterstock **page 80:** TrotzOlga/Shutterstock **page 81:** Feng Yu/Shutterstock **page 82:** (top) Neil Mitchell/Shutterstock (bottom) gnohz/Shutterstock **page 83:** (top) Lisa Turay/Shutterstock (bottom) tab62/Shutterstock **page 84:** rebvt/Shutterstock **page 85:** Shebeko/Shutterstock **page 87:** (top) marilyn barbone/Shutterstock (bottom) mommyknows/Flickr **page 88:** Daniel Loretto/Shutterstock **page 89:** Ispace/Shutterstock **page 92:** angelakatharina/Shutterstock **page 93:** tarog/Shutterstock **page 94:** (top) anweber/Shutterstock (bottom) Sarah/Flickr **page 95:** (left) Shebeko/Shutterstock (right) cakersandco/Flickr

SECTION III

page 96 (clockwise from top): Arno van Dulmer/Shutterstock, Mircea BEZERGHEANU/Shutterstock, Kokhanchikov/Shutterstock, w.g.design/Shutterstock **page 97 (clockwise from top left):** Al Mueller/Shutterstock, M. Cornelius/Shutterstock, Fotokostic/Shutterstock **page 98:** rimira/Shutterstock **page 99:** Alison Hancock/Shutterstock **pages 100-101:** Edsel Little/Flickr **page 102:** JGade/Shutterstock **page 103:** Steve Byland/Shutterstock **page 106:** Danny E Hooks/Shutterstock **page 107:** Kevin Krejci/Flickr **page 108:** kafka4prez/Flickr; **page 109 (both):** Timothy Musson/Flickr **page 110:** Kokhanchikov/Shutterstock **page 111:** (left) kafka4prez/Flickr (right) Timothy Musson/Flickr page 112: Tanis Saucier/Shutterstock **page 113:** normanack/Flickr **pages 114, 115, and 116 (both):** audaxl/Shutterstock **page 117:** Alison Hancock/Shutterstock **page 118:** Elena Blokhina/Shutterstock **page 119:** l i g h t p o e t/Shutterstock **page 120:** (bottom) Becky Sheridan/Shutterstock **page 121:** Katie Brady/Flickr **page 122:** sammydavisdog/Flickr **page 123:** sanddebeautheil/Shutterstock **page 128:** WilleeCole/Shutterstock **page 129:** Mircea BEZERGHEANU/Shutterstock **page 131:** Just chaos/Flickr **page 132:** Pete Markham/Flickr **page 133:** (top) clarence s lewis/Shutterstock (bottom) Ken Mayer/Flickr **page 134:** Zoran Vukmanov Simokov/Shutterstock **page 135:** Arno van Dulmer/Shutterstock **page 136:** (top) silversyrpher/Flickr (bottom) John Hritz/Flickr **page 137:** brewbooks/Flickr

SECTION IV

page 138 (clockwise from top): Cynthia Kidwell/Shutterstock, Ingrid Balabanova/Shutterstock, Ingrid Balabanova/Shutterstock, Anne Kitzman/Shutterstock **page 139** (clockwise from top left): kukuruxa/Shutterstock, Elena Elisseeva/Shutterstock, Alison Hancock/Shut-

terstock **page 140:** Africa Studio/Shutterstock **page 141:** Konstantin Sutyagin/Shutterstock **page 142:** aquatic creature/Shutterstock **page 143:** (top) Natalia Wilson/Flickr (bottom) Gorilla/Shutterstock **page 144:** Monkey Business Images/Shutterstock **page 145:** (top) Linda Hughes/Shutterstock (bottom) Le Do/Shutterstock **page 146:** fotografaw/Shutterstock **page 147:** basel101658/Shutterstock **page 148:** (top) lladyjane/Shutterstock (bottom) paul prescott/Shutterstock **page 149:** Alison Hancock/Shutterstock page 150: (left) Dancing Fish/Shutterstock (right) johnbraid/Shutterstock **page 152:** kesipun/Shutterstock **page 153:** Candy Frangella/Shutterstock **page 154:** (top) Shutterschock/Shutterstock (bottom) LesPalenik/Shutterstock **page 156:** Juan Nel/Shutterstock **page 157:** Ingrid Balabanova/Shutterstock **page 158:** ningii/Shutterstock **page 159:** c.byatt-norman/Shutterstock **page 160:** patpitchaya/Shutterstock **page 161:** photo25th/Shutterstock **pages 162–165:** Noah Weinstein **page 166:** de2marco/Shutterstock **page 167:** aaleksander/Shutterstock **page 170:** Dani Simmonds/Shutterstock **page 171:** EdCorey/Shutterstock **page 172:** Alison Hancock/Shutterstock **page 173:** (top) Alison Hancock/Shutterstock (bottom) dcwcreations/Shutterstock **page 174:** Artography/Shutterstock **page 175:** sxpnz/Shutterstock **page 177:** (top) Colleen Taugher/Flickr (bottom) Kathryn Depew/Flickr **page 178:** Miramiska/Shutterstock **page 179:** KellyNelson/Shutterstock **page 180:** (top) Vlue/Shutterstock (bottom) Shenjun Zhang/Shutterstock **page 182:** Aaron Amat/Shutterstock **pages 183 and 184:** vallefrias/Shutterstock **page 185:** Michael Major/Shutterstock **page 190:** photosync/Shutterstock **page 191:** Lucas Cavalheiro/Shutterstock **page 192:** His and Hers Homesteading (http://hisandhershomesteading.wordpress.com) **page 194:** rimira/Shutterstock **page 195:** Eva Solo A/S (www.evasolo.com) **page 196:** Shannon Holman/Flickr **page 198:** Igor Sokolov (breeze)/Shutterstock **page 199:** Laura Stone/Shutterstock **pages 200 and 201:** RDPixelShop/Flickr

SECTION V

page 202 (clockwise from top): nikky1972/Shutterstock, Sarycheva Olesia/Shutterstock, PerseoMedusa/Shutterstock, Shebeko/Shutterstock **page 203 (clockwise from top left):** Arjuna Kodisinghe/Shutterstock, Chamille White/Shutterstock, Kharkin Vyacheslav/Shutterstock **page 204:** AnnalA/Shutterstock **page 205:** withGod/Shutterstock **page 206:** Sarycheva Olesia/Shutterstock **page 207:** (top) www.sarahkolb.com/Flickr (bottom) Marianne Madden/Flickr **page 208:** avers/Shutterstock **page 209:** (top) Kzenon/Shutterstock (bottom) Diana Taliun/Shutterstock **page 210:** (top) Denis and Yulia Pogostins/Shutterstock (bottom) Christian Jung/Shutterstock **page 211:** Vsevolodizotov/Shutterstock **page 212:** Ryan DeBerardinis/Shutterstock **page 213:** Alena Brozova/Shutterstock **page 216:** Gordon Bell/Shutterstock **page 217:** Judy Orietz/Shutterstock **page 218:** (top) Michelle Marsan/Shutterstock (bottom) Anna Sedneva/Shutterstock **page 219:** Jenn Huls/Shutterstock **page 220:** Michelle Paccione/Shutterstock

Index

A

American Beekeeping Federation 136
artificial lighting for plants............168, 212-15

B

baby's bottom cream 86
bagels 39
basil 208
beekeeping..134-37
beer brewing...42-45
bee-sting allergy.......................................136
beeswax...90
bird feeder..102-5
bonsai 218
bread baking 34-41

C

candle making..88-91
canning 60, 62
castile soap..83, 87
cedar shavings 95
cheese making...24-26
chicken coop 120-21
    building a 124
chickens 118-127
    dual-purpose..121
    egg production of....................121, 122-23
    housing for 120-21, 124-27
    meat...122
    sexed versus straight-run 122
    types of...121-22
chives...208
cilantro...209
cleansers 80-83
climbing plants 180
cloche 148, 149-50
cold frame...148, 149
    building a 150
community gardening 198-201
companion planting............................159 176
compost... 112-17
    "browns" and "greens" in..............114, 116
    chicken manure as 123
    cold...106, 114, 116
    fourteen-day (UC Berkeley) method 116
    hot 114, 116
    with worms 106, 108, 114
compost tea...111
compost tumbler...117
condiments 28-33
conservatory (for plant protection)............ 148
container gardening 158, 162-164, 168-169,
    176, 177, 196, 206-8
cordials...46-49

D

dairy products...22-27
dilly beans 64
dried fruit..66-69
dried herbs 211
    for fragrance.. 94
    in tea...54-59
dryer bag, lavender 94
durum flour...18
Dutch lights..148

E

English muffins..38
essential oils...83, 86

F

flour, grinding your own.........................36, 41
fluorescent lighting for plants 214
freezing produce 70
freezing weather
    and rain barrels......................................101
    and worm bins 109
fruit cordials 46-49
fruits
    drying...66-69
    freezing...70-73

G

garden
    outdoor..140-45
    pollinator... 152-55
    raised-bed 170-73
    windowsill..204-11
ginger ale...52
goats 128-133
    behavior of...131-32
    housing for 132-33
    milking 130, 132, 133
    types of 130
gravity-fed watering systems.................190-93
greenhouse...148
grains, grinding into flour 36, 41
grow lights 214

H

hanging planters 168
herbs
    drying 56, 211
    growing 204-11
    suggested for tea 54-59
honey bees 136
hoop house 148
hot bed 148
hothouse 148

I

insect repellent 87
irrigation systems 182-89

K

ketchup 30

L

lavender 92
    dryer bag 94
    sachet bag.. 95
lip balm 86
light exposure of garden.....................142-43
lighting for indoor plants 168, 212-15
liqueurs 48

M

marble and granite cleaner...........................83
margarita mix 53
mason bees 134-37
mayonnaise...31
meat stock..14
mixers 50-53
mosquito disks 101
mulching 145
mustard..31

N

nesting box, building a 125-27
nonpressurized watering system..........192-93
noodles..18-20

O

orangerie.. 148
oregano 209

P

parsley... 210
pasta making...16-21
pickling 60, 64
pizza garden 158
planters
    hanging 168
    make your own 160-65
    on a deck 156-59
    self-watering 194-97
    stacked................................... 174, 176
    vertical 166-69
pollination..................................136, 154-55
    recommended plants for.........................155
poured candles......................................90-91
preserving foods
    by canning 60, 62
    by freezing 70-73

by pickling 60, 64
    in root cellar 74-77
pressure canning 62
    with pickling 64
pressure-watering system 182-89

Q

quinine... 52

R

Raabe, Robert D. 116
rain barrels..98-101
    safety with 101
raised beds 170-73
"Rapid Composting Method, The" 116
ravioli 20
ravioli filling...20
rolled candles...90
rolls 36
root cellaring..74-77
rosemary...59, 210
row cover 148

S

sachet bags...95
sage 59, 211
sandwich bread 37
sauerkraut..65
season extension....................................146-151
seasonings...28-33
self-watering planters.........................194-97
semolina...18
skin-care products 84-87
soap 83, 87
soil quality..................................142-43, 172
soil testing...144
spices, grinding your own............................ 33
squirrels, and bird feeders.........................104
stainless steel cleaner 83
stock..12-15

T

tea 54-59
    bags 56, 57, 58
    herbal 56-59
    infusers...56-57
    strainers 56, 58
thyme 59, 211
tomato sauce 63
tonic 52
topiaries...216-21
    recommended plants for.......................219
tortellini..21
trellises... 178-81
tub and tile cleaner 83
tunnel (for plant protection) 148

U

US Food and Drug Administration 86

V

vanilla extract 32
vegetables
    freezing...70-73
    growing for recipes..............................158
    harvesting..159
    root cellaring 74-77
vegetable stock..14
vermicomposting............................106-11, 114
vertical growing 166-69

W

watering methods 143, 182-89, 190, 194
wax for candles 91
window cleaner...82
windowsill garden..................................204-11
    plant suggestions for.......................208-11
worm bins 106-11, 114

Y

yogurt...26-27